Atomic Cannons and Nuclear Weapons
A Mystery of the Korean War

BY
ARTHUR G. SHARP

History Publishing Company LLC
Palisades, NY 10964

Copyright © 2017 Arthur G. Sharpe

All Rights Reserved. No part of this publication may be reproduced, stored in a retrieval system, or transmitted in any form by any means-electronic, mechanical, photocopying, recording or otherwise without written permission from the History Publishing Company LLC.

ISBN 9781940773452

SAN:850-5942

Published in the United States by
History Publishing Company LLC
Palisades, NY 10964

Printed in the United States of America on acid free paper

First Edition

Contents

Dedication v

Acknowledgements vi

Eleven Significant Facts About The Use Of Nuclear Weapons In The Korean War vii

Introduction 1

Chapter 1 May "Annie" Never Fire In Anger 13

Chapter 2 The Korean War: A Turning Point In The Development And Use Of Nuclear Weapons 20

Chapter 3 We Don't Know How To Conduct A Nuclear War, But We Can Prepare For One 27

Chapter 4 Here, World, You Take Responsibility For "The Bomb" 34

Chapter 5 The Race Goes To The Swiftest To Develop More Powerful Weapons 42

Chapter 6 Everyone Has A Point—And A Tipping Point 47

Chapter 7 "Mac Attacks" Abound 58

Chapter 8 Truman May Be Gone, But He Did Not Take The Nuclear Weapons With Him 67

Chapter 9 The Birth And Development Of " Atomic Annie" 75

Chapter 10 Marines Learn That Radioactive Doesn't Mean Changing The Radio Station 84

Chapter 11 Operation Crossroads 92

Chapter 12 At Least We Have A Nuclear Cannon 97

Chapter 13 Ike Visits Korea 103

Chapter 14 Do We Have To Drop Another A-Bomb? 111

Chapter 15 Harry And "Mac" Differ On The Attack 119

Chapter 16 The Chinese Are Coming En Masse 125

Chapter 17 We Like "Ike," But We Are Tired
Of The War 131

Chapter 18 Ike And "Annie" Who? 136

Chapter 19 Fehrenbach's History Stands Out 141

Chapter 20 The Bombs Are Ready 146

Chapter 21 What May Be The Real Story 151

Chapter 22 Anchors And A-Bombs Aweigh 158

Chapter 23 How Nuclear Bombs Saved Lives 164

Chapter 24 Was "Annie" Twins? 170

Chapter 25 "I Heard From A Friend Who Heard
From A Friend…." 176

Chapter 26 240? 280? What's 40mm Among Friends? 183

Chapter 27 Not Much On Which To Build A Reputation 191

Appendix A Un Resolution 82 195

Appendix B Un Resolution 83 197

Appendix C 198

Appendix D 200

Appendix E 202

Appendix F 204

Sources 206

Author Bio 211

DEDICATION

To the millions of men and women from the 22 nations (including South Korea) that responded to the United Nations' call for help to repel communism from South Korea. Their willingness to fight for freedom led to the creation of a free, independent, and economically viable South Korea, and delivered a damaging blow to the Communists' effort to combine and rule the entire Korean peninsula under their oppressive form of government.

ACKNOWLEDGEMENTS

Thanks to the Korean War veterans who contributed their stories and explanations pertaining to the use (or non-use) of nuclear weapons in the Korean War, especially those firsthand accounts about the creation and possible employment of the world's first nuclear cannon in Korea.

ELEVEN SIGNIFICANT FACTS ABOUT THE USE OF NUCLEAR WEAPONS IN THE KOREAN WAR

1. President Truman threatened to use nuclear weapons against the Communists in Korea.
2. President Eisenhower also threatened to use nuclear weapons during the Korean War.
3. The U.S. Army had available a nuclear cannon that could-- and may--have been employed in Korea.
4. Nuclear weapons were shipped from the U.S. to Guam and Japan during the Korean War for possible employment.
5. The only injuries caused by nuclear weapons during the Korean War were incurred by American service members and civilians.
6. General MacArthur had a plan in place to deliver nuclear weapons on Communist targets.
7. U.S. allies were against the use of nuclear weapons in the Korean War.
8. U.S. Air Force crews flew practice runs over Korea to assess their capability for dropping nuclear weapons.
9. The chief role of nuclear weapons in Korea was as a form of psychological warfare.
10. Exactly who would drop nuclear weapons in Korea was a source of contention between the U.S. Navy and the U.S. Air Force.
11. At least two U.S. Navy aircraft carriers, Lake Champlain and Valley Forge, allegedly carried atomic weapons and the planes and pilots trained to drop them.

INTRODUCTION

> *"The value of nuclear weapons since the first—and only—atomic bombs ever used in combat were dropped over Hiroshima (August 6, 1945) and Nagasaki (August 9, 1945) Japan has been strictly psychological."* (The author)

As we observe annually the anniversary of the first two uses in history of atomic bombs, on 6 and 9 August 1945, we note that they were also the last times they were used—up to this point, at least. (The contents of this book suggest that observation may not be 100% accurate.) The two atomic bombs dropped by the United States over Hiroshima and Nagasaki, Japan hastened the end of WWII. There have been wars since then, and numerous countries have developed nuclear weapons in the interim.** But, not one has employed them--at least that we know about. That may be due in part to decisions about their use made by U.S. political and military leaders in the "Forgotten War," aka the Korean War (1950-53).

** (Just for the record, there is a thin line between the terms atomic and nuclear weapons. According to "The Naked Scientists:"

"There are two major types of nuclear weapon/bomb: atomic and hydrogen bombs. An atomic bomb works by splitting large atomic nuclei (fission) such as uranium, or more usually plutonium. This releases quite a lot of energy. There is, however, a limited size you can make this, as the bomb tends to blow itself apart before it has all ignited.

Hydrogen bombs use an atomic bomb to ignite a nuclear fusion bomb, i.e., fusing hydrogen isotopes together to form helium. This releases a lot more energy, especially as the neutrons released by the fusion part of the bomb makes the fission part more efficient.")

We will use the terms nuclear and atomic interchangeably throughout this book, despite the technical differences.

Normally, historians don't associate the Korean War with nuclear weapons. There were none used during the war, so why would they? Again, maybe there were. But if there were, the story of their use will remain forever shrouded in a "mystery of history."

Even if nuclear weapons weren't used, they were at the forefront of political and military leaders' discussions as the war dragged on. Those leaders were not only Americans. After all, the war was technically a United Nations (UN) operation, a coalition involving 22 countries against North Korea, China, and Russia. Mainly, however, it was a U.S. effort.

Not surprisingly allies involved in the Korean War wanted a say in the use or non-use of nuclear weapons once the idea of their use was introduced, and they weren't shy about expressing their opinions, even though the U.S. was the only country that had them in its arsenal and ready to employ in 1950. (Russia didn't test its first nuclear bomb until 1949.)

The issue about employing nuclear weapons was on the table almost from the beginning of the war. But, the memories of the devastation the atomic bombs dropped on Japan lingered fresh in the leaders' and public's minds. So, the military and political poohbahs wrestled constantly with the moral and ethical issues of unleashing nuclear weapons once again, even as a way to achieve a total victory over the North Koreans, Chinese, and Russians—and Communism in general.

The questions they had to answer between 1950 and 1953 were complicated, just as they were in 1945. The only difference was that the world did not know what to expect in 1945. People had a better idea of the effects of nuclear weapons by the time the

Korean War started, and not all of them favored their use. That didn't make either President Truman's or President Eisenhower's decision to use or not use them any easier—not that either of them would have made it without advice.

(As Lee Iacocca pointed out, "I always go back to Harry Truman: Should we drop an atomic bomb to save 100,000 lives? That's a hell of a decision to make. Did he make that decision by himself? No, he had advisers.")

If they employed nuclear weapons, what would the rest of the world think? Did it matter what they thought? Where would they use them: just in North Korea, which had launched the initial attack against its southern brethren, or on Russia and China, North Korea's strongest allies? Who would pick the targets, and what would the targets be? Who would make the final decision to use nuclear weapons? How would nuclear weapons affect the environment? Would their use lead to an all-out worldwide race to develop bigger and more lethal weapons of mass destruction? Could the U.S. justify the costs of the nuclear weapons? Many of the questions are still unanswered today.

While leaders wrestled with these questions, the war dragged on. The two sides finally resolved their differences, albeit temporarily, before the final decisions re nuclear weapons were made. Those questions and answers are still pending, however. The Korean War is technically ongoing, since the two sides signed a cease fire, not a treaty, and both sides are armed with new and improved nuclear weapons. The three-year struggle ended in a stalemate, with troops of both sides dug in along the 38th Parallel, where it began. (Significantly, the South Koreans accepted the cease fire under duress, but they never signed any documents to acknowledge that fact.)

The North Koreans have violated that cease fire several times since 1953, and armed guards from South Korea and North Korea still glare at one another at the DMZ (demilitarized zone) in

Panmunjom, ready to shoot one another at the slightest provocation. The North Koreans have accused the UN of violating the agreement as well. Charges and counter-charges aside, technically, the war goes on, despite Eisenhower's efforts to end it then and subsequent attempts by various parties since then. Korea remains divided.

The Russians Are Cunning, The Russians Are Cunning

The underlying issue was the Communists' quest for control of a united Korea ostensibly under the aegis of the Soviet Union. In reality, the Communist country the U.S. feared the most was Russia, the least involved overtly in the war. But, the Russians' footprint on land and sea and in the air was all over the unpleasantness.

President Truman simply did not trust the Russians. He agreed with U.S. Secretary of State Dean Acheson, who told British Prime Minister Clement Atlee at a late 1950 meeting, "It had to be remembered that the central enemy [in Korea] was not China, but the Soviet Union. All the inspiration for the Korean action came from Moscow" (Harry Truman 397).

In a 15 March 1952 speech at the Columbia Scholastic Press Association, Truman declared, "We are not imperialists. We do not want any more territory. We do not want to conquer any people, or to dominate them. The Russian propaganda says that we are imperialists and want to conquer the world. That just isn't true. We know the Soviet Government is a menace to us and to all the free world. That is why we are building up our strength, not to march against them but to discourage them from marching against us and the free world." (The speech is available at
http://millercenter.org/president/truman/speeches/speech-3352.)

The Russians may not have fought personally against the U.S. in great numbers, but they interjected themselves vicariously through their allies.

It's Not Us; It's Those Damned North Koreans And Chinese

The North Koreans were using a considerable amount of Russian equipment when they crossed the 38th Parallel to attack South Korea, spearheaded by what many experts considered at that time to be the world's most powerful and unstoppable tank, the T-34.

There were two Russian anti-aircraft artillery divisions deployed well inside Korea, around Pyongyang. The Russians also maintained numerous control posts in theater, along with a searchlight regiment, a technical aviation division, two military hospitals, advisors, and assorted guards. Estimates suggest that 72,000 Russians troops were deployed inside Korea on a rotating basis during the war, 26,000 of them in 1952 alone. And they were supplying the Chinese troops after they entered the war in October 1950, albeit with somewhat outdated weapons.

U.S. sailors suspected Russian skullduggery at sea. One alleged instance included an attack on the destroyer USS Walke (DD-273) on 12 June 1951. Crew member QM2 Guy Willis recounted the story:

"In the summer of 1950, just after the Korean War started, I volunteered for the Navy again. (I had been in the Navy at age 17 and 18 at the end of WWII). I was immediately assigned to USS Walke (DD-723) in San Diego. Our assignment was to take her from the reserve fleet, get ready for combat, re-commission her and get to Korea. In short order we did just that. All under the excellent command and guidance of Captain Marshall Thompson, a 1936 graduate of the U.S. Naval Academy.

"On January 2, 1951, we left for Korea. Our assignment was to protect the carrier fleet off Wonsan, Korea. But often we were detached from that duty to go close in and fire our 5-inch guns at targets on land. We frequently experienced return fire from coastal batteries.

"At 7:40 a.m. on June 12, 1951, we were at sea as a part of the carrier screen when Walke experienced an explosion in the berthing compartments on the portside aft. It opened a 40-foot hole in her hull below and above the water line. Twenty-six shipmates never knew what (killed) hit them. Ten bodies were blown into the sea through that hole and never recovered. In addition, 40 shipmates were wounded.

The official report was that Walke struck a free floating mine. The North Koreans were known to have released these Chinese-made mines off their coast so that they would drift south on the prevailing current and into the operating area of the UN/US fleet off Wonsan. Many had been sighted and detonated by rifle fire from several ships.

However, at the time Walke was hit, other destroyers in the fleet detected submarine activity in the area. Several detached and dropped depth charges on a possible target. Later inspection of the damage to Walke's hull indicated it could have been caused by a mine or a torpedo.

At that time, neither the North Korean or Chinese navies had a submarine capability. In that area, only Russia did. Perhaps that is why the official report indicates the damage was done by a mine. Tanks, mines, rockets...the Russians were connected in one way or another to all of them."

It's All In Your Heads

U.S. Army Captain Leonard Kleckner, a specialist in psychological warfare, recalled a story of "cocky" Chinese prisoners of war who were boasting about a new rocket the Russians had provided them that had the potential to turn the war in their favor.

"We were catching rockets along the I Corps front, and while they weren't doing much damage, a few prisoners we picked up were cocky as hell—seemed to think they had a secret weapon from the Russians," Kleckner explained.

"It was the old Katusha, an obsolete rocket launcher the Russians used in WWII. We printed a leaflet showing we knew all about their secret weapon and had passed it up as old-fashioned. We said the Russians were peddling junk they couldn't use in exchange for good Chinese grain and cash," he continued.

"The reaction was sensational," Kleckner concluded. "In one week, the Chinese GIs gave their political commissars such a bad time the rockets disappeared from I Corps and have never showed up to any degree since" (Collier's 16-17).

That was one instance in which psychological warfare reached down to the lowliest privates in an army. That was not always the case as leaders threatened one another with weapons of one type or another, e.g., nuclear bombs. Unfortunately for allied personnel, not all the equipment the Russians supplied was inferior and old fashioned.

Those Russian Pilots Were Aces

Russian pilots, sometimes disguised as Chinese, flew technologically advanced (for the time) MiG-15s in aerial combat against UN aircraft. It is estimated that by 27 July 1953, the date of the cease fire, three-quarters of the aerial combat over Korea was between Russia and U.S. pilots. Furthermore, "Russian pilots flew 63,229 sorties during the war, compared with 22,300 flights by Chinese and North Korean pilots" (Wetterhahn 70-72). Yet, the Russians tried to keep their involvement in the war secret. Why?

One Russian combat pilot, Col. Yevgeniy Pepelyayev, intimated that one of the reasons was Stalin's fear of a nuclear attack on Russia by the U.S. Stalin took seriously Truman's implied threat to use nuclear weapons against the Communists. So, Pepelyayev said, "Stalin tried to conceal Russian participation by requiring pilots to remain behind a line drawn between Wonsan and Pyongyang in central North Korea" (Wetterhahn 74-75).

Secrecy or not, Russian pilots took a major toll on UN resources.

The Russians claimed to have shot down more than 1,300 U.S. planes during the war, in addition to a few Australian and South African aircraft. The numbers may be exaggerated, but even one allied plane shot down by a Russian fighter proves that Russia was involved overtly in the Korean War. That justifies the Americans' perception of Russia as their chief competitor for world power after WWII, as a National Security Council report known as NCS-68 stated.

NCS-68 Hits The Russian Nail On The Head

Two months before the North Koreans invaded the south, President Truman received NSC-68, "A Report To The National Security Council by the Executive Secretary on the United States Objectives and Programs for National Security (April 12, 1950)." A couple excerpts highlight how wary the U.S. government was of Russia's rise to power globally, especially as a rival nuclear power.

"The Soviet Union, unlike previous aspirants to hegemony, is animated by a new fanatic faith, antithetical to our own, and seeks to impose its absolute authority over the rest of the world. Conflict has, therefore, become endemic, and is waged, on the part of the Soviet Union, by violent or non-violent methods in accordance with the dictates of expediency.

"With the development of increasingly terrifying weapons of mass destruction, every individual faces the ever-present possibility of annihilation should the conflict enter the phase of total war....The Kremlin regards the United States as the only major threat to the achievement of its fundamental design....In a shrinking world, which now faces the threat of atomic warfare, it is not an adequate objective merely to seek to check the Soviet design, for the absence of order among nations is becoming less and less tolerable.

"This fact imposes on us, in our interests, the responsibility of world leadership."

The report included a lengthy section of the growing importance of atomic weapons and Russia's race to develop them. The opening paragraphs of Section VIII reveal that the U.S. was focusing all its attention on one enemy. As it turned out, the country's political leaders were placing all their eggs in the wrong basket, since it was North Korea that initiated the next war, not Russia, albeit it with Russian backing.

It is interesting to note that according to the report U.S. military and political leaders realized that nuclear weapons alone would not lead to total victory. They were considered just one part of a larger arsenal:

A. MILITARY EVALUATION OF U.S. AND USSR ATOMIC CAPABILITIES

1. The United States now has an atomic capability, including both numbers and deliverability, estimated to be adequate, if effectively utilized, to deliver a serious blow against the war-making capacity of the USSR. It is doubted whether such a blow, even if it resulted in the complete destruction of the contemplated target systems, would cause the USSR to sue for terms or prevent Soviet forces from occupying Western Europe against such ground resistance as could presently be mobilized. A very serious initial blow could, however, so reduce the capabilities of the USSR to supply and equip its military organization and its civilian population as to give the United States the prospect of developing a general military superiority in a war of long duration.
2. As the atomic capability of the USSR increases, it will have an increased ability to hit at our atomic bases and installations and thus seriously hamper the ability of the United States to carry out an attack such as that outlined above. It is quite possible that in the near future the USSR will have a sufficient number of atomic bombs and

a sufficient deliverability to raise a question whether Britain with its present inadequate air defense could be relied upon as an advance base from which a major portion of the U.S. attack could be launched.

It is estimated that, within the next four years, the USSR will attain the capability of seriously damaging vital centers of the United States, provided it strikes a surprise blow and provided further that the blow is opposed by no more effective opposition than we now have programmed. Such a blow could so seriously damage the United States as to greatly reduce its superiority in economic potential.

Effective opposition to this Soviet capability will require among other measures greatly increased air warning systems, air defenses, and vigorous development and implementation of a civilian defense program which has been thoroughly integrated with the military defense systems.

In time the atomic capability of the USSR can be expected to grow to a point where, given surprise and no more effective opposition than we now have programmed, the possibility of a decisive initial attack cannot be excluded."

Despite the NSC study's bold predictions, fortunately, or perhaps unfortunately, no American military or political leader was willing to okay the use of nuclear weapons in Korea, even with history on their side. After all, nuclear weapons had encouraged Japan to seek an end to a war. Results aside, perhaps it is that history that deterred the Americans from using their nuclear weapons in Korea, and spared the world from their use for at least the next three-quarters of a century.

Nuclear weapons did not mean only air-dropped bombs. The U.S. Army was developing an atomic cannon to complement a new generation of air-dropped weapons, one of which was a hoped for "game changer" called the hydrogen bomb. Experts believed that the development of the "H-bomb" meant that nuclear weapons were here to stay, as an excerpt from a 29 November 1952 editorial in The Nation averred:

"Now that the US has exploded its first hydrogen bomb, a negotiated peace with the Soviet Union is more important than ever.

"The announcement that the atomic age has now given birth to the H-bomb, said to be a thousand times more destructive than the 1945 A-bomb, must be considered in the light of several major realizations. On October 26, 1952, John Foster Dulles, President-elect and former WWII Allied Supreme Commander Dwight David ("Ike") Eisenhower's new Secretary of State, and Dr. Arthur H. Compton [a key figure in the Manhattan Project that developed the first nuclear weapons], in interviews with Richard G. Baumhoff of the St. Louis Port-Dispatch, agreed that it is now too late to outlaw or abandon the use of atomic weapons."

The nuclear race was already in full swing when NSC-68 was released, despite the lethal results of the atomic bombs dropped on Japan. The world saw the horrors of nuclear weapons when "Little Boy" and "Fat Man" were dropped over Hiroshima and Nagasaki respectively in 1945. And, compared to the "bigger and better" nuclear weapons the U.S. was developing in the lead-up to and during the Korean War, the first two bombs were small. That, combined with the fact that Russia was beginning its own nuclear weapons program in 1949, made world leaders take pause and deplore the use of such weapons in any war at any place post WWII. That did not stop them from threatening to use them.

So, when U.S. leaders mulled the use of nuclear weapons in Korea to bring the Communists to the peace talks table, they had to consider domestic and world opinion. Their choice would not only have an impact on the Korean War but on the future of nuclear weapons in general. Another excerpt from The Nation's editorial highlights that idea:

"Scientific advances in the field of atomic energy have outpaced political thought in the last five years and greatly complicated the immediate military problem. Our Atlantic allies will insist now more than ever that the scale and tempo of rearmament can be safely reduced. The explosion of the first British atomic-bomb on October 3 will encourage this belief. At the same time, the realization that we are more vulnerable than the Soviet Union to atomic attack—since American industry is more concentrated than Russian—will add to our uneasiness, and this uneasiness will become acute indeed when the Russians announce, as they will eventually, that they too have the H-bomb. In the meantime, our possession of the H-bomb will encourage the preventive-war enthusiasts to believe that now is the time to force a showdown."

It is obvious based on studies like NSC-68 and editorials like The Nation's that the use of nuclear weapons was a major concern to military and political leaders in the late 1940s and during the Korean War. Yet, the choice regarding the use of nuclear weapons in that war is rarely considered in any depth by historians. To many of them it is no more than a footnote to a bit of unpleasantness that was actually the second nuclear weapons war.

The results of "Fat Man" and "Little Boy" were fresh in the minds of the world in the 1950-53 timeframe, and they played a role in the ultimate decision by American leaders not to use nuclear weapons in Korea, except as an implied threat. This book examines the role of nuclear weapons in the Korean War and their impact on other nations' decisions not to use them in any war thereafter as anything but threats.

Those threats have not become reality since 1953, at least not yet. That distinction between threat and reality has become an overlooked part of the legacy of the Korean War—and the premise of this book.

Chapter 1
May "Annie" Never Fire In Anger

> *"The taboo on the use of nuclear weapons in limited wars—indeed the very notion of a 'limited' war itself—had not yet taken root [in 1950]; the Korean war defined these principles, but there was little reason to expect, when it broke out, that its conduct would reflect them."* John Lewis Gaddis

Nuclear weapons played a significant role in the outcome of the 1950-53 part of the Korean War, even though they were never used. How, then, did nuclear weapons, including an innovative-for-the-time "atomic cannon," fit into the war and the American strategy to end it? And why all the mystery about their use, especially the cannon known as "Atomic Annie?"

U.S. Air Force veteran and former TSgt George ("Ski") Sherman uses the following story to inform JROTC and ROTC unit members in his area about the role of the little-known piece of "atomic" artillery that played a central role in bringing relative peace to Korea after three years of fighting.

"Prior to May 25, 1953, the only way to deliver an atomic device to the enemy was by airplane. Other delivery systems, such as rockets and missiles, were still in various stages of planning and development. The U.S. Army insisted that, since its soldiers were required to take ground in a battle, they needed a tactical atomic weapon to enhance their chances."

The problem, according to Sherman, was that the Air Force argued for control of all nuclear armaments. In fact, many U.S.

Army leaders in the Korean War era feared that the army's future was in jeopardy. Major-General John Singlaub, USA (ret) advanced that viewpoint in his book, Hazardous Duty:

"Within the Defense Department, it was the Air Force that benefited most from our massive retaliation policy. They had the planes and missiles to deliver the nuclear weapons on China or the Soviet Union. The more massive our threat, the Soviet (and later Chinese) counter threat, and our own anti-counter threat response became, the greater the Air Force's need for manned strategic bombers, fighter-bombers, and long-range missiles. By 1954, there were even rumblings that the Air Force would eventually replace the Army, which was viewed in some quarters as being on the verge of obsolescence. Conventional ground forces, it was argued, simply couldn't be defended against atom bombs. And nuclear weapons were, after all, cheaper than a huge standing army. Even though we had a peacetime draft, expanding the U.S. Army to the size needed to realistically counter the combined threats of the huge Soviet and Chinese ground forces would have completely altered our peacetime society. So for almost a dozen years--until the Cuban Missile Crisis of 1962--America relied increasingly on its nuclear strike force, to the detriment of the Army."

Sherman expanded that theme:

"The largest tactical artillery pieces that were readily available at this time were the 240mm Howitzer and the 8-inch guns of WWII," he noted. "An entirely new weapon and its ground equipment were needed. The weapon should be a long-range artillery gun piece, capable of being moved cross country, around corners of most village roads, and not too heavy for most bridges. Its firing tube or barrel would have to be large enough to handle the smallest atomic projectile then in development."

That new weapon was the 280mm cannon, which came to be known as "Atomic Annie" or 'Atomic Ike.'

"The 280mm (11 inch) shell was designated the T-124,"

Sherman continued. "That shell had a Mark 9 nuclear warhead. The complete shell was 54-1/2 inches long, and weighed 803 pounds. It had three-timed altitude fuse settings to choose from.

"During 1949-1950, the Joint Chiefs of Staff (JCS), knowing that the Army's Picatinny Arsenal in Dover, NJ was close to completing the T-124 Atomic Shell, authorized the Army's ordnance department to produce twenty guns that could be used successfully. A new era in Army and U.S. history was about to emerge.

"The big gun's tubes were made at the Army's Watertown, NY Arsenal; the gun was assembled at its arsenal at Watervliet, NY. By early 1952, 'Atomic Annie,' as the gun was nicknamed, was ready. The gun tube was 40 feet, 2 inches long. The complete unit, ready for traveling at a speed of 35 miles per hour, was 84 feet, 2 inches long.

"The gun and its carriage were moved by two transporters specially made to move the gun. The transporters were tractors, one at either end of the assembly, which could either pull or push. They could even move the weapon sideways by turning at right angles to the center section. The 'A' and 'B' units were also connected by telephone or radio.

"The transporters carried the crew and equipment and possessed the capability to carry two non-nuclear rounds. The complete unit, road ready, weighed 85 tons, which was not much more than the heaviest artillery piece then in use. The unit, which could fit into a landing ship if it was necessary, could be fired from a lanyard (a strong cord used to activate a system) at twenty feet or electrically from a position miles away.

"The system had its own electrical generator and hydraulic systems for lifting the gun onto its traveling carriages and ramming home the shell power charges. The turntable allowed the gun to rotate a full 360 degrees with a four-man crew. It was also capable of being prepared by manual power.

"On the arrival of 'Atomic Annie' for duty at the artillery center at Fort Sill, Oklahoma, a full year was spent training with the gun testing various techniques (with non-nuclear) warheads on the Fort Sill [Oklahoma] Firing Range.

"On May 25, 1953, at Camp Desert Rock, a desolate outpost in the Frenchman Flats area of the Nevada Proving Grounds for atomic weapons, U.S. Army Observers waited anxiously to find out if a nuclear war head would stand being suddenly hurled at a speed of 2,060 feet per second, furiously spinning, and exploding in an air burst over a target 6-1/2 miles away. Or, would it explode prematurely in the cannon barrel, obliterating the immediate surroundings with its explosive force of 15,000 tons of TNT, equal to that of 'Little Boy,' the atomic bomb that wasted Hiroshima in 1945?

"The ultimate test shot was code named 'Grable' by the Atomic Energy Commission (AEC). LtCol Donald L. Harrison wrote, 'I had the task of physically ramming out the atomic round in case of a misfire. I gave a big sigh of relief when the activated round cleared the big tube with only the big gun's normal thunderous noise.'

"Undoubtedly, the rest of the gun crew gave a collective sigh of relief when the atomic round went off, exploding as planned more than 6 miles away at a height of 524 feet. Its radioactive mushroom-shaped cloud of earth and dust rose majestically 35,000 feet into the atmosphere.

"For the firing of the atomic shell, the gun was on ground high enough to be in a direct line of sight to ground zero and the explosion. Once the firing data was computed, the gun crew, positioned in a trench nearby, received firing commands by telephone. The gun was fired by a 20-foot lanyard.

"The next step for 'Atomic Annie' and her nineteen brothers and sisters was for combat gun crews and field artillery units to be organized and shipped out to support U.S. ground forces. "Atomic

Annie" was assigned to the 265th F.A.B.N. The 265th was deployed to Baumholder, Germany in November 1953."

This is where another Korean War veteran, Dick Payne, picks up the story.

"After serving as a Navy Hospital Corpsman in Korea with the 1st Marine Division in 1953, I was assigned to Baker Medical Company, 2nd Medical Battalion, 2nd Marine Division, Camp Lejeune, North Carolina. While I was there, the battalion Chief Petty Officer told me he had a special duty assignment for me.

"I was to take a field ambulance with a Marine driver and report to the U.S. Navy Port Commander at a certain pier in Wilmington, North Carolina. He said he did not know what it was about, but we were not to tell anybody where we were going. We were to stay until we were told to return.

"I was instructed also to prepare a medical kit with supplies for possible accidents—splints, battle dressings, etc.—including morphine syrettes, which Corpsmen normally did not carry in the U.S. I was also told to draw a .45 pistol with spare magazines from the armory.

"The Marine ambulance driver was told to take his M-1 rifle with bayonet and several bandoliers of ammo. We were ordered to take a week's supply of C-rations and given a cash advance so we had money. We received permission to sleep in the ambulance and to bring back receipts for any money we spent.

"As we rode to Wilmington, the ambulance driver and I speculated about what we were getting ourselves into. I knew it was very unusual for a Hospital Corpsman to carry a loaded weapon in the U.S.

"We found the pier and reported to the naval officer in charge. He asked what we were doing there, so I relayed what little we had been told. He said they were loading a U.S. Army atomic cannon artillery unit for duty in Europe—the first to be shipped there. And, he related, since Camp Lejeune was the closest military base with a large medical facility, as a courtesy, they had notified the base because there was the potential for accidents—atomic and otherwise—and problems with protesters against the use of atomic weapons.

"He noted that he had not requested any direct assistance, but since we were there we might as well stay until the ship sailed.

"We stood by and watched the ship being loaded. I wondered what I could do if there was an atomic accident. I would probably be vaporized like everybody in the dock area. As to protesters, the Marine ambulance driver and I seemed to be the only ones with loaded weapons. Were we expected to hold off any angry crowds that might burst through the gates?

"I do not remember how long we were there, but no accidents occurred and we saw no protesters, although they may have been kept outside the gates of the pier.

"As the dock lines were unfastened, and the ship prepared to sail, the naval officer waved good-by to us and we returned to Camp Lejeune. Thus ended my small part in the history of atomic warfare."

Sherman explained what happened once "Annie" reached Germany.

"In the three years 'Annie' served in Germany, she participated in three maneuvers, five demonstrations, three 600-mile road trips, and two dignitary visits. In 1956, Cardinal Spellman of New York City gave a blessing that 'she never be fired in anger, but that she would ever stand for peace with her power.'

"The 280mm cannon was gradually taken out of service during the 1960s. 'Atomic Annie' now rests in retirement at Fort Sill, Oklahoma, while 'Atomic Ike' can be found on display at the Army Ordnance Museum at Aberdeen, MD.

"I had the proud distinction of serving on 'Annie' from March 1953 until 1956 from Fort Sill to Fort Bragg to

Baumholder, Germany. As I understand it, one gun was sent to Korea from one of the battalions in Okinawa."

Originally, news of the deployment was only hearsay. Evidently, one was actually sent to Korea, although that has never been verified. In any event, the "Atomic Annie" story in particular and atomic weapons in general is a piece of Korean War and Cold War history that has been largely ignored and overlooked in today's history.

NOTE: Sherman's story appeared in The Graybeards, Jan/Feb 2011, p. 15. Payne's account was in The Graybeards, Mar/April 2011, p. 73. The Graybeards is the bimonthly 80-page publication of the Korean War Veterans Association.

Chapter 2
The Korean War: A Turning Point In The Development And Use Of Nuclear Weapons

"For some reason I do not fear the Atom Bomb era that is with the world. There are and will be, I am confident, still enough good men and women in this world to control properly the various advancements of men." Leonard P. Schultz

On 25 June 1950 North Korean military forces poured across the 38th Parallel, the dividing line between North and South Korea, in an effort to unite the two countries under a Communist government by force. The well-armed, well-trained, Russian-backed North Koreans were successful initially. They pushed their poorly armed, insufficiently trained, American-backed South Korean foes almost to the southern limits of the Korean Peninsula.

Almost immediately, the United Nations (UN) condemned the North Korean aggression via Resolution 84 (see Appendix A). It did not, however, order military aid to the South Koreans. That came two days later with the passage of Resolution 83 (see Appendix B). The UN forces, composed primarily of U.S. units, entered the war and stemmed the North Korean onslaught temporarily, but not immediately

U.S. involvement was under the umbrella of the United Nations. However, the way the command structure was stacked and the war was conducted it was clear that the United States was in charge of running it. The numbers bear that out: nearly 6,000,000 Americans served in the military in some capacity between 1950 and 1953, not all of them in Korea.

The UN and North Korean forces seesawed back and forth as 1950 drew to a close. In October of that year the Chinese "volunteer" armies entered the war to help their North Korean allies. Until late February 1951, that returned the advantage to the Communists. Then, the UN's first decisive victory in the war, at a place called Chipyong-ni, turned the tide of battle once again.

"About Phase"

By early 1951, the UN forces had grown to a point where they were able to fight the Communists on an equal basis. The war became a stalemate, as both sides dug in around the 38th Parallel. How static the war became is summed up in a history of the Korean War published by the U.S. Army that identifies thirteen phases in the ground fighting:

1950

- Withdrawal to the Pusan Perimeter (June 25 - July 31)
- Defense of the Pusan Perimeter (August 1 - September 14)
- UN Counteroffensive (September 15 - November 24)
- Withdrawal from the Yalu River (November 25 - December 31)

1951 (the Yalu River was the dividing line between North Korea and China)

- Enemy High Tide (January 1 - 24, 1951)
- Attack and Counterattack (January 25 - February 28, 1951)
- Crossing the 38th Parallel (March 1 - April 21, 1951)
- Enemy Strikes Back (April 22 - May 19, 1951)
- UN Resumes Advance (May 20 - June 24, 1951)
- Lull and Flare-up (June 25 - November 12, 1951)

1952

- Stalemate (November 12, 1951 - June 30, 1952)
- Outpost Battles (July 1 - December 31, 1952)

1953

- The Last Battle (January 1 - July 27, 1953)

The back and forth between the two sides suggests that neither could budge the other. They both sought an edge that would give them the upper hand in the war. For the Communists it was sheer numbers and psychological warfare, at which they excelled. They believed they could overwhelm the UN forces by launching mass assaults on them with overwhelming numbers of troops. The UN counterpunched with air strikes and advanced weapons such as smart bombs and napalm. Both combatants' tactics worked to some extent, but they did not break the stalemate.

Let's Introduce Freud And Adler To Korea

The U.S. complemented its combat efforts with psychological warfare and threats of different types, including nuclear weapons, which only the U.S. possessed at the time. (Russia, the undeclared ally of the Chinese and North Koreans, was in the process of developing its nuclear weapons program by the late 1940s.) The nuclear weapons became part of the psychological part of the war, which is defined as "[communicating] the ideas and information intended to affect the beliefs, emotions and actions of the enemy in order to lower his morale, destroy his will to fight and to induce him to take action beneficial to our cause." Eventually, the Communists became inured to the threats of the nuclear weapons because they became empty words.

U.S. military leaders were aware that words alone could not win a war. Colonel Kenneth Hansen, chief of the Far East

Command's Psychological Warfare Section during the Korean War, stressed that point: "Nobody in this shop thinks we can win this war with just words. Propaganda is a weapon, like tanks or planes or artillery. But you can't win with tanks *or* planes alone, and words without something to back them up are—well, just words" (Collier's 15).

Secretary of the Army Frank Pace seconded that. "The art of applying psychological as well as physical force against a military opponent has become an accepted element of modern warfare. As a support weapon, psychological warfare has taken its place with the tank, the gun and the airplane. Its mission is to reduce the cost in man power and materiel necessary to obtain an objective. It is here to stay" (Collier's 15).

The psychological warfare Hansen and Pace envisioned did not cut costs or materiel needs in Korea. And, as far as threats of nuclear weapons were concerned, they were all addressed to the Communist nations' leaders, not to the troops in the field. All the troops were subjected to from a psychological standpoint was a barrage of propaganda leaflets and radio broadcasts urging them to surrender. Even they were of limited value.

In general, the warfighters were not concerned about the use or non-use of nuclear weapons in Korea. The Communist soldiers could only wonder if atomic bombs could be any worse than the "fire bombs" and napalm dropped in large quantities by allied bombers and fighter planes. They had a better chance of being struck directly by a 25-pound bundle of 10,000 leaflets dropped from a C-47 than they did of succumbing to a nuclear blast. The only difference was that conventional bombs were more destructive.

(Early in the war leaflets and bombs were combined. The 1st Radio Broadcast & Leaflet Company dropped leaflets over areas about to be bombed as a warning to civilians to "get out of Dodge." No doubt the enemy troops in the area could read those same leaflets and evacuate the area as well. Whether that same warning courtesy would have been afforded to people about to be subjected to nuclear weapons was not known.)

By mid-September 1950 almost every town and village in North Korea had been destroyed or damaged severely by allied bombing raids which included conventional and incendiary bombs. The total destruction proved to be a blessing in the long run for the North Koreans and a bane for the allies. The devastation eliminated possible targets for nuclear weapons later in the war, which lessened the leverage the U.S. had regarding its threats of using them. Unless the U.S. intended to use them on Russia and China, the threat of nuclear weapons in North Korea was practically laughable.

Even If We Use Nuclear Bombs, Where Do We Use Them?

The U.S. threats to use nuclear weapons sounded serious. There was no doubt that crews were trained to deliver them. Former U.S. Air Force B-50 tail gunner Len Johnson trained at Castle Air Force Base in California in 1951 "to deliver an A-bomb to Russia if needed." The major question was where to drop them. That created a conundrum for U.S. political and military leaders.

To start, for the most part nuclear bombs were of little value as anti-troop weapons in Korea. The front was spread so wide that nuclear weapons would not have had much effect on small concentrations of the enemy. In addition, opposing troops were so near one another at times, practically close enough to shake hands, that nuclear weapons could have done as much damage to UN forces as they did to the enemy.

Second, there were no major metropolitan areas such as Hiroshima and Nagasaki to destroy. Pyongyang, the capital, was the only city in North Korea with a population exceeding 100,000 people. (Its population was 500,000.) The U.S. Air Force and Navy had done such an effective job of decimating North Korea's cities

Atomic Cannons and Nuclear Weapons

and infrastructure targets that a nuclear bomb or two wouldn't be particularly effective.

Third, employing nuclear weapons outside North Korea, in either China or Russia, would be political suicide. Doing so would not only anger people across the globe, but it might lead to World War III, which was the last thing UN and U.S. leaders wanted. It appeared that their value was strictly psychological. About all the U.S. military planners could do was speculate about the damage nuclear weapons could cause—and that was after they lost opportunities to use them.

Coulda, Shoulda, Woulda...

One opportunity in particular was during the UN withdrawal from North Korea in late 1950 after MacArthur's aborted attempt to reach the Yalu River in force. (Only a few allied troops actually reached the Yalu River, and they withdrew rather quickly.) That was when the Chinese entered the war and surprised the UN troops heading north. The Chinese relentlessly pushed the UN troops back toward the 38th Parallel. Only UN air attacks kept the retreating ground forces from annihilation.

Far East Air Force (FEAF) administrators estimated that their airmen killed or wounded 40,000 Chinese troops during the withdrawal, the equivalent of five divisions. Army researchers revealed later that if FEAF had used nuclear weapons they would have inflicted at least twice as many casualties on the Chinese.

The Army study concluded that if FEAF air crews had dropped nuclear weapons on the Chinese they "could have taken a horrible toll of enemy troops." The estimates suggested that one 40-kiloton air burst weapon exploded over the dense enemy concentration at Taechon on the night of 24/25 November 1950 would have destroyed some 15,000 of 22,000 troops (Futrell 655). The report also said that six nuclear bombs of the same magnitude dropped over the 95,000 massed Communist troops in the

Pyonggang-Chorwon-Kumwha area between 27 and 29 December 1950 might have killed at least half of them.

The numbers got larger as the report grew longer. The researchers estimated that if FEAF had dropped six 30-kiloton nuclear bombs on the 70,000 to 100,000 enemy troops lined up north of the Imjin River on 31 December 1950 in preparation for an attack against the 8th Army, they might have killed 28,000 to 40,000 of them.

Finally, on 7-8 January 1951, 18,000 North Korean troops gathered across from Wonju to launch an attack against UN forces. The study claimed that if FEAF had dropped only two 40-kiloton bombs, between 6,000 and 9,000 enemy troops would have died. Alas, no bombs were dropped on any of these enemy troop formations. All the results included in the study were mere speculation, which sums up the entire history of nuclear weapons use in the Korean War.

Okay, why didn't the U.S. approve the use of nuclear weapons in these cases? The answers are the same as would apply to other instances during the war when UN air crews could have used them. First, there may not have been enough trained pilots and bombardiers to drop the nuclear weapons effectively. Second, in most places UN and enemy troops were so close together that friendly forces in large numbers may have died alongside their Chinese and North Korean foes.

Finally—and this is what renders the study moot—UN forces did not have any idea of the numbers of enemy forces gathered or where specifically they were positioned. There simply was not enough intelligence available to supply that information. Therefore, the numbers suggested by the Army study are of historical interest, but they were based on something that did not happen. The UN lost its chance to inflict severe damage on the enemy with nuclear weapons, which was the case from January 1951 to the end of the war.

Chapter 3
We Don't Know How To Conduct A Nuclear War, But We Can Prepare For One

"I expressed the hope that we would never have to use such a thing against any enemy because I disliked seeing the United States take the lead in introducing into war something as horrible and destructive as this new weapon was described to be." (Dwight D. Eisenhower)

Even though it wasn't clear exactly how the U.S. could wage an atomic war in Korea, since the action was limited to "trench warfare" after the first few chaotic months, the U.S. Air Force Strategic Air Command (SAC) was prepared to conduct one, which led to the only casualties resulting from the use of nuclear weapons during the Korean War.

"On August 1, 1950, the decision was made to send the 9th Bomb Wing to Guam as an atomic task force immediately. Ten B-29s, loaded with unarmed atomic bombs, set out for the Pacific.

"On August 5, one of the planes crashed during takeoff from Fairfield-Suisun Air Force base near San Francisco, killing a dozen people and scattering the mildly radioactive uranium of the bomb's tamper around the airfield. The other planes reached Guam where they went on standby duty (pbs.org). (The "tamper" was a neutron reflector in an atomic bomb that delays the expansion of the exploding material, making possible a longer-lasting, more energetic, and more efficient explosion.) It is interesting to note that the article describing the crash uses the words "highly explosive," not atomic or nuclear. (Read an account of the crash in Appendix C.)

A Flight Fraught With Fear

One of the problems with nuclear weapons had always been their instability. Thus, they had to be armed in flight or transported in parts. In WWII Captain William ("Deak") Parsons, USN, who helped research, develop, and assemble the atomic bomb, flew aboard the Enola Gay en route to Hiroshima to facilitate the drop. He crawled into the bomb bay to arm the bomb, which had to be armed in flight because of its unstable design. A similar problem existed with the nuclear weapons being transported from the Fairfield-Suisun base to Guam.

The bombs had to be transported in two parts and in secrecy. The dense uranium core and the high explosive outer casing would be carried to Guam in separate planes. The routes and times of the flights were not revealed. One B-29 that left Fairfield-Suisun on 5 August 1950 crashed almost immediately after take-off. It was carrying the high-explosive portion of a Mark IV bomb.

Here is an account from the official U.S. Air Force report:

"On 5 August 1950, B-29, SN 44-87651, crashed, burned, and exploded 5 minutes after takeoff from Fairfield-Suisun AFB, CA, causing fatal injuries to 12 crewmen and passengers. Eight crewmen and passengers received minor injuries. Extensive damage to private and government property and injuries to both civilian and military personnel were caused by a subsequent explosion of the bomb on the aircraft.

"The pilot, Captain Eugene Q. Steffes, was at the controls, with Brigadier General Robert F. Travis acting in command pilot capacity. At 2200 PST, the aircraft was cleared for takeoff on runway 21 left, which is 8,000 feet long. The wind was 17 knots from the southwest. A full power check (2,800 ROM and 48 inches) was made, and the brakes were released for takeoff.

"Just prior to liftoff, the number two engine propeller malfunctioned, and the aircraft commander ordered the number two propeller be feathered. After liftoff, the pilot actuated the gear switch to the up position, and the gear did not retract. Due to the increased drag (feathered number two engine and the lowered gear), the rising terrain ahead and to the left, and the inability of the aircraft to climb, the aircraft commander elected to make a 180-degree turn to the right back toward the base.

"Upon completion of the turn, the left wing became difficult to hold up. The aircraft commander allowed the aircraft to slide to the left to avoid a trailer court. A crash landing was imminent as the altitude of the aircraft was only a few feet above the ground. The aircraft struck the ground with the left wing down at approximately 120 mph. All ten people in the rear compartment were fatally injured. General Travis and one passenger in the forward compartment received fatal injuries; all other crewmembers and passengers escaped with only minor injuries" (Check-Six.com).

That wasn't the worst of it, though. The bomb aboard did not explode when the plane hit the ground. About twenty minutes later the high explosives in the bomb casing ignited. That resulted in an explosion that was felt and heard over thirty miles away. The blast caused severe damage to a nearby trailer park on the base. That added to the total of dead and injured people, which amounted to 180: 7 killed, 49 admitted to hospitals, and 124 wounded superficially.

Ironically, the dead and wounded were the only casualties resulting from the use of nuclear weapons in the Korean War, despite all the threats made to use them. And, they occurred on U.S. soil, far away from the Korean battleground. Such are the fortunes of war.

Just What We Need: A Fire Near Our Atomic Bomb

The rest of the bombs reached Guam safely. The fact that SAC positioned them for use in Korea demonstrates that the U.S. wanted them accessible, even if it wasn't going to employ them. It also shows that atomic weapons played a part in the conduct of the war albeit, as it turned out, a limited one. But, at least one atomic bomb was shipped from Guam to Japan, as former Navy officer Frank Barron recalls:

"This event occurred in the spring of 1953. I was a young officer stationed aboard USS

Eversole (DD 789), a Gearing Class destroyer stationed out of Long Beach, California. My job at this point was the Electrical Officer. I subsequently became Chief Engineer and stayed on the ship for another three years.

"The Korean War was beginning to wind down, but North Korea would not come to the peace talks table in a serious fashion. General Eisenhower had been installed as president in February 1953. He said nothing was off the table and insinuated that the atomic bomb could or would be used if necessary.

"We were in Yokosuka, Japan, on the southeast coast of Japan, when we received an urgent query to see if we could get underway and how fast. (Yokosuka was one of the major bases for the Japanese navy during WWII). Our commanding officer, Captain Victor Delano, indicated we could move within 2 to 3 hours.

"We were directed to go to Guam and escort an ammunition ship, USS Chara (AKA 68), back to Yokosuka. It took us a couple days to get to Guam, where we stayed overnight. We set out the next morning, escorting Chara. The standard procedure in those days was to be stationed

2,000 yards directly ahead of the ship which was being protected, theoretically at least, from submarines.

"There was a designated zigzag course which we both followed in an attempt to confuse the submarines. One morning, about the second day, we were told to take station immediately forty miles ahead. This was a strange request, so the captain contacted Chara. The Tactical Commander on the ammo ship, Captain Babb, was asked to repeat the instructions.

"They were repeated: 'Take station immediately 40 miles ahead.'

"Within a few hours the order came back: 'Resume normal steaming and course.'

"Everyone on the ship was really curious what that was all about. When we got back to Yokosuka it happened that the commanding officer of the ammunition ship was a friend of Delano. They had known each other before, so they met at the Officers Club. The captain of the ammunition ship told Delano that he had escorted an atomic bomb.

"I did not learn some of this story until I attended an Eversole reunion in Colorado Springs [Colorado] in 1985, some 30+ years later. What had happened was there was a small fire on the ammunition ship. Of course, there was the danger that it might trigger an explosion. They thought we would be far better off at a distance, and that's why we were sent forty miles ahead. "The story stuck in my mind. Remember, this is the idea of escorting the only atomic bomb with the intention of using it since World War II.

"In 2014 I happened to be at the Farmers Market in Rome, Georgia, where I ran into a friend of mine of many years, Buddy Andrews. The subject came up of being in the military. He said he had been in the Navy and was on the ship that carried an atomic bomb to Japan to be possibly used in Korea.

"I asked Buddy if by any chance they had a fire on board. He said, 'Yes.' They had one in the boat locker. He remembered that every man on his ship ran to put it out, because they all knew about the atomic bomb aboard.

"I said, 'Buddy, you'll never believe this. I was on board the ship escorting you from Guam to Yokosuka. We have known each other for seventy years or more, and I never knew that you were there.'

"As I reviewed the incident, I came to the conclusion that we carried on board my ship about 300 men and about 20 officers. The crew on his ship probably comprised 100 men and maybe 7 officers. We are talking about 420 people in the whole United States. The odds that two of them were from the same town was incredibly unusual."

So was the fact that this little-known tidbit of history ever came to light. Nobody would have learned about it by reading Chara's generic Korean War history, which disclosed only that the ship carried ammunition:

"With the outbreak of the Korean War, Chara was transferred to Service Force, Pacific Fleet, for duty as an ammunition ship, transporting and transferring all types of ammunition at sea to fleet units. She cleared San Francisco 16 September 1950 to replenish TF 77 and support the evacuations of Hungnam and Wonsan before returning to San Francisco for overhaul 26 March 1951. In her second Korean tour, 19 July 1951 to 18 May 1952, she joined the Mobile Logistics Support Force in operations in the Wonsan-Songjin bomb-line triangle, and in emergency lifts of Korean POWs from Ko.je-do to Ulsan. Another tour of providing at-sea replenishment of ammunition preceded the end of hostilities." (Source: hazegray.org/danfs/amphib/axa58.htm)

The same holds true for Eversole. Its history is just as generic:

"The second Eversole was launched 8 January 1946 by Todd-Pacific Shipyards, Inc., Seattle, Wash.; sponsored by Mrs. S. R. Eversole, mother of Lieutenant (junior grade) Eversole; and commissioned 10 May 1946, Commander B. P. Ross in command.

Atomic Cannons and Nuclear Weapons

[The first Eversole, DE 404), was torpedoed on 28 October 1944 by the Japanese submarine I-45. Its entire crew was killed or wounded.]

"Eversole arrived at San Diego, her home port, 6 October 1946, and in the years prior to the Korean War, twice sailed to the Far East for duty with the 7th Fleet, patrolling off China and Japan. She sailed from San Diego 1 May 1950 for another such tour, and thus was in the Orient upon the opening of the war. Until 8 February 1951, when she returned to San Diego, she screened the fast carrier task forces as they launched air strikes against North Korean targets.

"During her second tour of duty in the Korean War, from 27 August 1951 to 10 April 1952, Eversole bombarded Hungnam, Wonsan, and other points along the east coast of Korea, and served in the Blockading and Escort Force, with ships of the navies of Great Britain, Canada, Netherlands, Australia, New Zealand, and the Republic of Korea. From 17 November 1952 to 29 June 1953, she served a similar tour of duty." (Source:hazegray.org/danfs/destroy/dd789txt.htm)

It is ironic that the only way people around the world may have ever known about the role of atomic bombs in Korea would have been if the crew aboard Chara had been unable to distinguish that small fire and they and their ship had blown up. As it was, the story did not come to light until 60+ years later in a round-about way. That is how much history is revealed: through stories long hidden and related by unexpected sources.

Chapter 4
Here, World, You Take Responsibility For "The Bomb"

"We face an enemy whom we cannot hope to impress by words, however eloquent, but only by deeds-- executed under circumstances of our own choosing."
Dwight D. Eisenhower

One reason the role of nuclear weapons in the Korean War is often overlooked or underemphasized by historians, journalists, teachers, foreign policy experts et al might be because the weapons were never used other than as propaganda and psychological weapons during it. But the threat of their use was always in the background. And, they played a part in President Harry Truman's decision to fire General Douglas MacArthur from his position of commander of UN military forces in Korea. That, too, is often forgotten or ignored. Strangely enough, Truman did not mention nuclear weapons in his 11 April 1951 radio address to the nation, nine months after the U.S. had entered the war.

The president began his talk by explaining why the U.S./UN was in Korea:

"I want to talk to you plainly tonight about what we are doing in Korea and about our policy in the Far East.

"In the simplest terms, what we are doing in Korea is this: we are trying to prevent a third world war.

"I think most people in this country recognized that fact last June. And they warmly supported the decision of the Government to help the Republic of Korea against the Communist aggressors.

Now, many persons, even some who applauded our decision to defend Korea, have forgotten the basic reason for our action.

"It is right for us to be in Korea now. It was right last June. It is right today."

He, like so many of his contemporaries and successors, blamed the Korean War on the Soviet Union, suggesting that the Communists wanted to "run all Asia from the Kremlin." He explained that "It was the Soviet Union that trained and equipped the North Koreans for aggression. The Chinese Communists massed 44 well-trained and well-equipped divisions on the Korean frontier. These were the troops they threw into battle when the North Korean Communists were beaten.

"The question we have had to face is whether the Communist plan of conquest can be stopped without a general war. Our Government and other countries associated with us in the United Nations believe that the best chance of stopping it without a general war is to meet the attack in Korea and defeat it there.

"That is what we have been doing. It is a difficult and bitter task."

He made only one reference in the address to bombs, but in general terms:

"We do not want to see the conflict in Korea extended. We are trying to prevent a world war—not to start one. And the best way to do that is to make it plain that we and the other free countries will continue to resist the attack.

"But you may ask why we can't take other steps to punish the aggressor. Why don't we bomb Manchuria and China itself? Why don't we assist the Chinese Nationalist troops to land on the mainland of China? [NOTE: That was a question Ike would have to deal with in terms of nuclear weapons in 1954 and 1958.]

"If we were to do these things we would be running a very grave risk of starting a general war. If that were to happen, we would have brought about the exact situation we are trying to prevent.

"If we were to do these things, we would become entangled in a vast conflict on the continent of Asia and our task would become immeasurably more difficult all over the world."

That was as far as he went regarding tactics and strategy. The implication—at least in hindsight—was that he did not want to worry Americans about another atomic bomb attack or the possible adverse effects, especially since this time the enemy might have had the capability to respond in kind. (Read the speech at http://millercenter.org/president/truman/speeches/speech-3351.)

News About The A-Bomb Is A Dud To Ike

The U.S. became the first country in the world to use atomic bombs when it dropped two over Japanese cities (Hiroshima on 6 August 1945 and Nagasaki on 9 August 1945) in an attempt to hasten the end of World War II. One--almost the only--American military leader was dead-set against the use of the bomb. That was General Dwight David Eisenhower (Smith 450).

Ike learned about plans to drop atomic bombs at a July 1945 conference at Potsdam, Germany. Secretary of War Henry L. Stimson, a member of the "Top Policy Group" which controlled the Manhattan Project and the man who reported to the president on its progress, told Ike about the successful tests of the bomb. The general was less than thrilled at the news.

Ike had two major objections to the use of atomic weapons: 1) Japan was already defeated and the public relations fallout would be detrimental to the U.S. "I thought that our country should avoid shocking world opinion by the use of a weapon whose employment was no longer mandatory as a measure to save American lives," he told Stimson. 2) He was a realist about the use of atomic weapons, recognizing that once the first one was employed it would more than likely not be the last. He and Truman thought alike in that respect.

Michael Korda highlighted Ike's ambivalence: "Neither Ike nor MacArthur viewed the atomic bomb as an 'unthinkable' weapon. Ike was fond of pointing out that every major advance in weaponry had once been described as 'unthinkable'--the repeating rifle, the machine gun, gas, tank, the bomber--but 'all weapons in due course become conventional weapons,' as he would tell Winston Churchill. And the atomic bomb was, in any case, not an 'unthinkable' weapon; it had been used, against Ike's advice, on the Japanese, and arguably had brought about their surrender...That is not to say that Ike wanted to use atomic bombs, any more than MacArthur did" (Korda 652).

Ike's ambivalent position about atomic bombs and his opposition to their use might have been unique at the time, but it did not change in future wars or situations in which he was offered the opportunity as president to employ nuclear weapons. Ironically, Korea was not one of them.

Here's Your Chance: Use The Bomb

Historians report that in 1954 Ike was given two chances to use nuclear weapons in war situations: in Vietnam and during the Formosa Strait crisis (Smith 450). A third scenario involving the possible use of nuclear weapons, again in the ongoing Formosa Strait area, presented itself in 1958. In all three cases Ike said no.

One offer was put on the table during the struggle between the Chinese Communists and Chinese Nationalists for political dominance in China.

"Despite warnings from the U.S. against any attacks on the Republic of China, five days before the signing of the Manila pact, the PLA (People's Liberation Army) unleashed a heavy artillery bombardment of Kinmen on September 3, and intensified its actions in November by bombing the Tachen Islands. This renewed Cold War fears of Communist expansion in Asia at a time when the PRC (People's Republic of China) was not recognized by the United States Department of State. Chiang Kai-shek's government was supported by the United States because the ROC (Republic of China) was part of the containment of communism which stretched from a devastated South Korea to an increasingly divided Southeast Asia.

"On September 12, the U.S. Joint Chiefs of Staff recommended the use of nuclear weapons against mainland China. Eisenhower, however, resisted pressure to use nuclear weapons or involve American troops in the conflict. However, on December 2, 1954, the United States and the ROC agreed to the Sino-American Mutual Defense Treaty, which did not apply to islands along the Chinese mainland. This treaty was ratified by the U.S. Senate on February 9, 1955." (Source: https://en.wikipedia.org/wiki/First_Taiwan_Strait_Crisis)

In Vietnam, U.S. Secretary of State John Foster Dulles allegedly offered to supply atomic bombs to the French in 1954 to help them save their troops at Dien Bien Phu, but they did not accept (BBC). The plan was called "Operation Vulture."

The plan included as many as 98 B-29s from Okinawa and the Philippines that would drop 1,400 tons of bombs on positions held by the Viet Minh. Another version of the plan envisioned sending 60 B-29s from U.S. bases in the region, supported by as many as 150 fighters launched from U.S. Seventh Fleet carriers, to bomb Giap's positions. [Note: General Vo Nguyen Giap was described in an October 4, 2013 New York Times obituary as "the relentless and charismatic North Vietnamese general whose campaigns drove both France and the United States out of Vietnam."]

It also included an option to use up to three small atomic weapons on the Viet Minh positions in support of the French. The Joint Chiefs of Staff drew up plans to deploy tactical atomic weapons, U.S. carriers sailed to the Tonkin gulf, and reconnaissance flights over Dien Bien Phu were conducted during the negotiations.

Admiral Arthur W. Radford, the Chairman of the U.S. Joint Chiefs of Staff and top American military officer, gave this nuclear option his backing. US B-29s, B-36s, and B-47s could have executed a nuclear strike, as could carrier aircraft from the Seventh Fleet. (Source: https://en.wikipedia.org/wiki/Operation_Vulture)

Decision against the operation

Ike's vice president, Richard M. Nixon, seemed keen for a fight in Vietnam, and even suggested that the U.S. might have to send troops there. Ike disagreed with both Nixon's willingness to employ U.S. troops and the use of nuclear weapons, but he did consider participation in Vietnam if the British agreed to join the U.S. They declined.

Besides, Ike did not believe that air power alone, whether or not it included nuclear weapons, would win the burgeoning war. Moreover, he did not think the French Air Force was capable of carrying out nuclear attacks, and he was reluctant to assign the mission to U.S. pilots. Ike also considered the adverse political risks of such attacks. He just did not see any benefits in authorizing them. As a result, he opted not to intervene.

The third time Ike declined to use nuclear weapons was in 1958, again during the ongoing Formosa Strait (Taiwan) crisis:

"On the day after the Chinese began shelling the Quemoy islands on August 23, 1958, U.S. Air Force Headquarters apparently assured Pacific Air Forces 'that, assuming presidential approval, any Communist assault upon the offshore islands would trigger immediate nuclear retaliation.'" Yet President Dwight D. Eisenhower rejected the use of nuclear weapons immediately, even if China invaded the islands, and emphasized that under no circumstances would these weapons be used without his approval.

"Caution against nuclear use didn't mean not planning for it, however, and in the years after the Taiwan Strait crisis an enormous nuclear build-up occurred in the Far East. The numbers started to decline in the 1970s, and for a period during the 1980s and first half of the 1990s, nuclear planning against China was reduced to reserve force contingencies. In the past decade, however, China has again become a focus for U.S. nuclear strike planning (Nukes)."

Since Eisenhower demonstrated three times that he was against nuclear warfare, it may be safe to conclude that he really was not serious about using atomic bombs in Korea or anywhere else. That idea was made clear by historians Stephen E. Ambrose and Tom Wicker.

Ambrose noted that "Eisenhower agreed with [L. Robert Oppenheimer] that an atomic arms race was madness; he also believed that if the American people were told, in graphic detail, of the destructive power of the H-bomb, they would support him in any genuine disarmament proposal (Ambrose 339)."

Wicker also alluded to Ike's antipathy towards atomic weapons.

"In this situation Eisenhower supposedly threatened the North Koreans and the Chinese with atomic weapons, whereupon they hastily came to terms. In fact, Eisenhower only allowed his Communist adversaries to *fear* that he might use the A-bomb--in 1953 a reasonable apprehension that he never denied. Thus, he kept his options open, as he preferred. He never issued a public and probably did not make a private threat; indeed, it's hard to see how an atomic attack would have differed--except in its greater severity--from the major offensive he did not want to launch, or why the former would not have provoked the Soviets and the latter might have (Wicker 27)."

Potato, potahto; tomato, tomahto ...whether Truman or Eisenhower ever directly threatened to use nuclear weapons in Korea or merely implied that they would is a moot point. The fact is that neither of them authorized the use of nuclear weapons there.

If they had, it might have set a precedent for their use in future wars. What they did, however, was open a global discussion on the use or threats to use nuclear weapons, which continues to this day.

Chapter 5
The Race Goes To The Swiftest To Develop More Powerful Weapons

"We need to be clear that there will not be many great atomic wars for us, nor for our institutions. It is important that there not be one. We need to liberate our own great resources to shape our destiny." (L. Robert Oppenheimer)

Americans faced some critical questions after World War II regarding their nuclear weapons development programs, e.g., do we continue to develop nuclear weapons, drop our program, maintain our nuclear weapons but place them under a third party such as the UN...? The U.S. attempted to go the third route right after the war.

The U.S. government appointed a five-man commission on 7 January 1946, only five months after the atomic bombs were dropped in Japan and four months after Japan surrendered, to look into the feasibility of turning control of nuclear weapons to the UN. The committee comprised a cross-section of distinguished members who knew their way around nuclear energy/weapons and the politics associated with their use:

- Dean Acheson, Chairman: Under Secretary of State (later Secretary of State during the Korean War)
- James Conant, a chemist who had worked on the development of synthetic rubber and on the Manhattan Project, which developed the first atomic bombs
- Vannevar Bush, an engineer, inventor, and science

administrator, and head of the U.S. Office of Scientific Research. He was the director of the Office of Scientific Research and Development, which controlled the Manhattan Project
- John McCloy, a "Man for All Seasons," who was an advisor to every U.S. president from Franklin Delano Roosevelt through Reagan
- General Leslie R. Groves, who had been the military officer in charge of the Manhattan Project

The Commission And The UN Are On Different Pages

In the letter of transmittal leading its report, also known as the Acheson–Lilienthal Report or Plan, the committee wrote:

"Anticipating favorable reaction by the United Nations Organization on the proposal for the establishment of a committee to establish the problems arising as to the control of atomic energy and other weapons of possible mass destruction, the Secretary of State has appointed a committee of five members to study the subject of controls and safeguards necessary to protect this Government so that the persons hereafter selected to represent the United States on the Commission can have the benefit of the study."

Another key paragraph read:

"In particular, we are impressed by the great advantages of an international agency with affirmative powers and functions coupled with powers of inspection and supervision in contrast to any agency with merely police-like powers attempting to cope with national agencies otherwise restrained only by a commitment to 'outlaw' the use of atomic energy for war. In our judgment the latter type of organization offers little hope of achieving the security and safeguards we are seeking."

The commission's anticipation went unrealized. As it turned out, the UN wasn't interested in taking control of weapons that could potentially destroy the world. Not surprisingly, one of the primary opponents of international control was Russia, which realized that the U.S. could not monopolize the possession of nuclear weapons forever—or even for the short haul—since it was developing its own.

When Bernard Baruch, the U.S. representative to the United Nations Atomic Energy Commission, proposed his June 1946 plan for international control of atomic energy, Russia rejected it immediately. The Russians protested that it was unfair, since the U.S. already had nuclear weapons and they didn't. They made a counter-offer: the U.S. should dispose of its nuclear weapons. Then, they claimed, international controls and inspections would make sense.

The U.S. was not about to do that. A stalemate ensued, which lasted until the Korean War began. That left the U.S. to rely on the Atomic Energy Commission (AEC), which Congress signed into law on 1 August 1946.

The essence of the law said that the U.S. should pursue and control the development of "atomic" science and technology for peaceful and defense purposes. Paradoxically, Congress declared that atomic energy "should be employed not only in the Nation's defense, but also to promote world peace, improve the public welfare, and strengthen free competition in private enterprise." How atomic bombs could at one time be both defensive weapons and promote world peace was a question for philosophers to resolve.

There was a considerable amount of debate as the law establishing the AEC was in the works. The central question was who would control atomic energy. Ultimately, a collection of politicians, military planners, and scientists decided that it should be under civilian control. That did not mean that the pursuit of nuclear weapons would be placed on a back burner. Seven years later, President Eisenhower renewed the call for international control in his 19 April 1953 "Cross of Iron" speech delivered to the American Society of Newspaper Editors. He advocated for:

1. "1. The limitation, by absolute numbers or by an agreed international ratio, of the sizes of the military and security forces of all nations.
2. A commitment by all nations to set an agreed limit upon that proportion of total production of certain strategic materials to be devoted to military purposes.
3. International control of atomic energy to promote its use for peaceful purposes only and to insure the prohibition of atomic weapons.
4. A limitation or prohibition of other categories of weapons of great destructiveness.
5. The enforcement of all these agreed limitations and prohibitions by adequate safeguards, including a practical system of inspection under the United Nations.

The details of such disarmament programs are manifestly critical and complex. Neither the United States nor any other nation can properly claim to possess a perfect, immutable formula. But the formula matters less than the faith—the good faith without which no formula can work justly and effectively.

The fruit of success in all these tasks would present the world with the greatest task, and the greatest opportunity, of all. It is this: the dedication of the energies, the resources, and the imaginations of all peaceful nations to a new kind of war. This would be a declared total war, not upon any human enemy but upon the brute forces of poverty and need.

The peace we seek, rounded upon decent trust and cooperative effort among nations, can be fortified, not by weapons of war but by wheat and by cotton, by milk and by wool, by meat and by timber and by rice. These are words that translate into every language on earth. These are needs that challenge this world in arms." (Read the speech at http://millercenter.org/president/eisenhower/speeches/speech-3357)

The proposals were a bit paradoxical considering that he offered them three months before the UN and the Communists signed a truce in Korea. He was threatening the use of nuclear weapons in the Korean War at the same time he advocated banning them altogether or limiting their use. Whether or not the mixed messages were intentional to confuse the enemy is anybody's guess. In any event, Ike didn't have any more success in his quest for international control than did his predecessor.

Chapter 6
Everyone Has A Point—And A Tipping Point

> *"The fact that no limits exist to the destructiveness of this weapon [the 'Super', i.e. the hydrogen bomb] makes its very existence and the knowledge of its construction a danger to humanity as a whole. It is necessarily an evil thing considered in any light. For these reasons, we believe it important for the President of the United States to tell the American public and the world what we think is wrong on fundamental ethical principles to initiate the development of such a weapon."* Enrico Fermi

There were strong arguments on both sides of the debate regarding which way the U.S. should go with its nuclear weapons program. People who saw the devastating results of the two bombs dropped in WWII were appalled at the destruction and loss of life they caused. Military leaders decried the effects, but saw in nuclear weapons a deterrent to future wars. Ultimately, the U.S. opted in January 1950 to pursue and upgrade its nuclear weapons program, not only by building bigger and better bombs that could be dropped by planes, but with weapons like atomic cannons.

President Truman explained his decision by saying, "I've always believed that we should never use these weapons. I don't believe we ever will have to use them, but we have to go on making them because of the way the Russians are behaving. We have no other course."

The circular reasoning was akin to an old Czech saying:

"When a Czech owns a goat, his neighbor does not yearn for a goat of its own; he wants the neighbor's goat to die." Truman really did not want a "hydrogen bomb" of his own; he simply hoped the Russian's attempt to build theirs would die. And, if it didn't, well...we still had one of our own.

The "for every action there's an equal reaction" theory kicked in.

The Russians Rush Into Their Own Program

The Russians responded to the U.S.'s decision to accelerate its nuclear weapons development program by upgrading their own. That was an immediate concern to U.S. leaders, and something they watched carefully. There were no wars in progress following World War II and none anticipated, which they thought gave them ample time to address the pros and cons of nuclear warfare. They did not foresee getting involved with another war anytime in the near future, since all their enemies had been vanquished after WWII—or so they thought. The Korean War caught them by surprise.

The Korean War offered the first opportunity for any country to use nuclear weapons in the field, and the only country that had them available in any number was the U.S. There were plenty of political and military leaders advocating their use. There were also many who did not.

Regardless of individuals' views, military and political leaders thought they had plenty of time to settle the issue without any enemy to interfere with their debate.

The one possible enemy American leaders could identify was Russia, which did not have a nuclear weapons capability in the late 1940s. Hardly anyone reckoned that only five years after WWII ended a relatively unknown country named Korea would be the "host" of the next war—and the subject of discussions about the use of nuclear weapons.

Bigger And Better Bombs By The Barrelful

A look at the development of nuclear weapons between the end of WWII and the start of and during the Korean War reveals that the U.S. had hundreds of them in its arsenal. No one knew how many there were exactly. President Truman noted that "In no document in my office, in the AEC, or anywhere in government, could anyone find the exact figure or the number of bombs in stockpile, or the number of bombs to be produced, or the amount of material scheduled for production" (Harry Truman 302).

Whether anyone was keeping count or not, the nuclear bombs got more powerful and diverse as each "generation" was rolled out. So did the conventional bombs such as Razons, the first bombs equipped with flight control surfaces, and Tarzons (upgraded Razons). The use of such bigger bombs became more prevalent as the war progressed.

Chart of Strategic Nuclear Bombs

The descriptions below identify all the Strategic Nuclear Bombs designed to be carried by aircraft before and during the Korean War. It does not include those carried by various missiles, nor the many nuclear shells designed for use by Army artillery. The list demonstrates that the U.S. had a plethora of options from which to choose for whatever purpose if it decided to use nuclear weapons during the Korean War.

Mk-1

Yield:15-16 Kt (Kilotons)
Fusing:Airburst
Number Produced:5
Weight:8,900 lbs
Dimensions:28" x 120"

The "Little Boy," the atomic bomb dropped on Hiroshima. It was a gun-assembly HEU bomb. Never stockpiled, only five assemblies were completed. They were retired by November 1950.

Mk-3

Yield:18-49 Kt
Fusing:Airburst
Number Produced:120
Weight:10,300 lbs.
Dimensions:60" x 128"

This is the "Fat Man" atomic bomb, dropped on Nagasaki, Japan. It was a Plutonium implosion bomb. The basic design was modified and upgraded over the next ten years. 120 were produced between April 1947 and April 1949. They were all retired in late 1950, which precluded their use in Korea.

Mk-4

Yield:1-32 Kt
Fusing:Airburst
Number Produced:550
Weight:10,800-10,900 lbs.
Dimensions:60" x 128"

The Mark 4 was a redesign of the "Fat Man." It was the first assembly-line produced nuclear bomb. It could be configured for various yields – 1, 3.5, 8, 14, 21, 22, and 31 kilotons.

Mk-5

Yield:6-120 Kt
Fusing:Airburst or Contact
Number Produced:140
Weight:3,025-3,125 lbs.
Dimensions:43.75" x 129" or 132"

The Mark 5 represented a major breakthrough in nuclear bomb design. The high efficiency implosion bomb could be configured to yield three times the explosive power of the Fat Man. Although it was about the same length, it had a thinner body.

Equally, if not more important, it weighed less than a third of the Fat Man. It was later used as the primary, or first stage, in the first thermonuclear devices. It could be used to yield 6, 16, 55, 60, 100, or 120 kilotons. It entered the operational stockpile in June 1952. The last one was retired in January of 1963.

Mk-6

Yield:8-160 Kt
Fusing:Airburst or Contact
Number Produced:1,100
Weight:7,600-8,500 lbs.
Dimensions:61" x 128"

The Mark 6 was an improved high-yield and lightweight Mk-4. It could be configured to yield 8, 26, 80, 154, or 160 kilotons. It was produced from June 1951 through early 1955. It was retired in 1962.

Mk-7

Yield:8-61 Kt
Fusing:Airburst or Contact
Number Produced:1,700-1,800
Weight:1,645-1,700 lbs.
Dimensions:61" x 128"

The Mk-7 "Thor" was a light-weight, multipurpose tactical bomb. The 92 lens implosion system permitted yields of 8, 19, 22, 30, 31 or 61 kilotons. Over 1,700 were produced between July 1952 and February 1963. They were in service from July 1952 until 1967.

Mk-8

Yield:25-30 Kt
Fusing:Pyrotechnic Delay
Number Produced:40
Weight:3,210-3,500 lbs.

Dimensions:14.5" x 116" or 132"

Nicknamed "Elsie," the Mark 8 was the first gun-assembly HEU bomb produced since the "Little Boy." It was designed as an earth-penetration weapon that could be used against hardened targets, such as underground command posts. Forty were produced between November 1951 and May 1953. It was retired in June 1957, replaced by the Mk-11.

Source: http://www.strategic-air-command.com/weapons/nuclear_bomb_chart.htm

Let's Develop A Real Bomb

The Mark series bombs were lethal, at least by WWII standards. But, the United States military still wanted something bigger. They got it: a thermonuclear, aka, hydrogen, bomb, which was approximately 1,000 times more powerful than conventional nuclear devices. All those "Mark" series bombs were like firecrackers compared to the hydrogen bomb.

On 1 November 1952, in the middle of the Korean War, the U.S. detonated its—and the world's—first thermonuclear weapon, the hydrogen bomb, on Eniwetok atoll in the Pacific. That gave the United States an advantage in the nuclear arms race with the Soviet Union, albeit temporarily. It also alarmed some of the people who recognized how dangerous a hydrogen bomb could be.

U.S. Senator Styles Bridges (R-NH), chairman of the Senate Appropriations Committee, wrote in a January 1954 article in Collier's Magazine:

"The hydrogen device which we set off November, 1952, on Eniwetok did something which was not adequately explained.

"It didn't smash or crumble things. It vaporized them. It didn't melt sand grains so that they ran together to form lava, as an atomic explosion would. It converted dirt, trees, rock and metal machinery to dust and gas. A substantial area composed of hard physical substances—lumber, earth, steel, copper, lead—was changed in about 20 seconds into atmosphere. Those who saw the test wondered what more scientists must do to convince mankind that an era has ended . . . and that a civilization will end unless men reconcile their differences peacefully.

"After Eniwetok, information on the destructive potential of hydrogen weapons began to reach Capitol Hill. One of the staggering statistics indicates that any thousand buildings—hotels, hospitals, warehouses, in San Francisco, Cleveland or Detroit—could be wiped out in one hydrogen blast. Detroit's real property improvements are valued at $10,000,000.000, its schools at one-half billion, churches at one-half billion; damage from an atomic raid over Detroit could reach $14,000,000,000. . . ."

Congress, which had been all but silent on the use of nuclear weapons during the Korean War, suddenly became interested in them right after it ended.

Predictably, Russia successfully tested a thermonuclear device in 1953. That explosion accelerated a nuclear weapons competition that culminated by the 1970s in a club of thermonuclear bomb owners that included seven countries. Not surprisingly, even some of the people who had helped develop the atomic bomb begged the U.S. not to build the hydrogen bomb. Chief among them were David Lilienthal and Robert Oppenheimer, whose reputation was dashed as a result.

Oppenheimer' Opposition Creates Calls Of "Communist"

Oppenheimer protested the development of the hydrogen bomb so vociferously that his detractors accused him of being a Communist, a spy, and a man who could not be trusted. William Borden, the former director of the Joint Congressional Committee on Atomic Energy, levied the charge of "spy" against Oppenheimer, without any evidence.

It was a sudden downfall for Oppenheimer. He had been appointed as chairman of the General Advisory Committee of the

AEC, which announced its opposition to the new bomb in October 1949. Four years later he was relieved of his secret nuclear research duties for the government and lost his security clearance when Ike ordered that he no longer be granted access to top-secret information pending an investigation into Borden's charges. (President Lyndon B. Johnson exonerated Oppenheimer in 1963.)

In a July 1953 article in the magazine Foreign Affairs, following Russia's hydrogen bomb test, Oppenheimer opined that an atomic arms race between and among nations could only result in disaster. (He included the United Kingdom as a third nuclear power.) And, he declared, such a race was inane. He asked whether it made any difference as to whether one country had 2,000 nuclear weapons or 20,000. Oppenheimer compared Russia and the United States to "two scorpions in a bottle, each capable of killing the other, but only at the risk of his own life."

He noted that "the atom, too, was given a simple role, and the policy followed was a fairly simple one. The role was to be one ingredient of a shield: a shield composed also in part of the great industrial power of America, and in part of the military and, even more, the political weaknesses of the Soviet Union. The rule for the atom was: "Let us keep ahead. Let us be sure that we are ahead of the enemy."

Oppenheimer also stressed that the American public should be made aware of the existence and lethality of nuclear weapons as a check on government's willingness to use them:

"As a first step, but a great one, we need the courage and the wisdom to make public at least what, in all reason, the enemy must now know: to describe in rough but authoritative and quantitative terms what the atomic armaments race is. It is not enough to say, as our government so often has, that we have made 'substantial progress.' When the American people are responsibly informed, we may not have solved, but we shall have a new freedom to face, some of the tough problems that are before us."

Atomic Cannons and Nuclear Weapons

Whether Oppenheimer's ideas were valid or not, they--and he--were not well accepted by many politicians at the time. Their dismissal led to a rapid loss of prestige for one of the U.S.'s top scientists, who had recently been a member of Project Vista, an effort designed to assess the feasibility of the use of atomic weapons in Korea.

Oppenheimer's situation was symptomatic of the U.S.'s paradoxical position on nuclear weapons in the Korean War. Political leaders realized that nuclear weapons were lethal, and that the more powerful they became the bigger a threat they were to humanity. Yet, they hoped that their mere existence and predictions of the destruction they could cause would serve as deterrents to using them.

They opted to maintain them in their arsenals and hope they would not have to use them. The Korean War provided them with an opportunity to practice restraint while threatening to use nuclear weapons. Meanwhile, they continued to develop state-of-the-art nuclear weapons and expanded the tools of delivery.

While new and improved nuclear weapons rolled off the assembly lines, the U.S. Army started a development program to build a cannon that could fire nuclear shells if need be. Its principal use would be for conventional weapons, but if it could be adapted to fire more lethal weapons, all the better. Nuclear bombs, nuclear artillery shells…the tipping point had been reached.

Russia Plays A Large Hand In The Korean War

Japan had ruled Korea with an iron fist from 1905 to 1945. The Japanese acquired it during a brief bit of unpleasantness with the Russians in 1905 known as the Russo-Japanese War. The dispute between the countries was resolved by U.S. President Theodore Roosevelt, for which he earned the Nobel Peace Prize.

The terms, which were outlined in the Portsmouth Treaty, gave Japan the right to govern Korea. The Russians were not particularly happy with the outcome, but they stayed out of Korea—until 1945. (As it turned out, the Koreans, who had no say in the decision either way, were less than enthralled by the decision to let the Japanese rule them.)

The Japanese governed a united Korea for forty years, and not always kindly. At the end of WWII negotiators removed Japan as the protector of Korea, and split the country into two parts. The Russians were granted responsibility for the northern part; the Americans took control of the southern part. The "two Koreas" were divided by the 38th Parallel. There were efforts by world leaders to create a united Korea, but the southern and northern Koreans could not agree on who would rule the entire country. Plus, their forms of government were diametrically opposed.

Meet "Orner" Rhee

The Russians and Chinese advocated a Communist government for the entire country and nominated Kim Il Sung as the leader. Sung's South Korean counterpart and his sponsors, the U.S. and Western Europe, demurred. They preferred their own president, Syngman Rhee, as the head of the proposed united country.

Rhee, a devoted non-Communist, whose first name might as well have been "Orner," welcomed that suggestion. He had a habit of doing things his own way, which sometimes upset UN military leaders and allied political poohbahs. That penchant lasted right up until the end of the fighting, when he refused to sign any cease fire that did not allow for a united Korea under his rule.

Il Sung and Rhee threatened to invade one another's territory to unite the country. Il Sung beat Rhee to the punch and almost succeeded, until the UN stepped in. A few months later, the Chinese joined the fray on the North Koreans' side.

UN and Communist forces bobbed and weaved like two drunken prizefighters for a few months. When it looked like the

UN forces commanded by General MacArthur were floundering in early 1951, President Truman called for a new commander. That ignited a firestorm of opinion pro and con. Truman got his way, and a major crisis erupted when he relieved MacArthur of his command on 11 April 1951.

Chapter 7
"Mac Attacks" Abound

"MacArthur could never see another sun, or even a moon for that matter, in the heavens, as long as HE was the sun." Peter Lyon

Historians generally provide reasons like MacArthur's insubordination, penchant for over-managing, old age, failure to recognize or care about Chinese intervention, his insistence on using atomic bombs in Korea, or a combination thereof to explain his dismissal as Commander in Chief of UN forces in Korea by President Truman in 1951. They cite as an example of his subordination a letter MacArthur wrote to the U.S. House of Representatives Joe Martin chastising President Truman for his failure to attract the Communists to the negotiating table.

As MacArthur's 20 March 1951 note to Martin suggests, he wanted complete victory. The implication was that Truman might not have been in the war to win it.

Letter to Representative Martin of Massachusetts:
(From Congressional Record of April 5, 1951)
20 March 1951

Dear Congressman Martin:
I am most grateful for your note of the eighth forwarding me a copy of your address of February 12. The latter I have read with much interest, and find that with the passage of years you have certainly lost none of your old time punch.

My views and recommendations with respect to the situation created by Red China's entry into war against us in Korea have been submitted to Washington in most complete detail. Generally these views are well known and clearly understood, as they follow the conventional pattern of meeting force with maximum counter force as we have never failed to do in the past. Your view with respect to the utilization of the Chinese forces on Formosa is in conflict with neither logic nor this tradition.

It seems strangely difficult for some to realize that here in Asia is where the Communist conspirators have elected to make their play for global conquest, and that we have joined the issue thus raised on the battlefield; that here we fight Europe's war with arms while the diplomats there still fight it with words; that if we lose the war to communism in Asia the fall of Europe is inevitable, win it and Europe most probably would avoid war and yet preserve freedom. As you point out, we must win. There is no substitute for victory. With renewed thanks and expressions of most cordial regard, I am,

Faithfully yours,
DOUGLAS MACARTHUR.

    The general was not making many friends in Washington DC with insinuations such as those contained in his note. Even his Army counterparts were unhappy with MacArthur.
    General Omar Bradley, the Army's chief of staff at the time, railed at MacArthur for suggesting that the U.S. should permit Chinese Nationalist leader General Chiang Kai-Shek to attack mainland China, where the Communists were headquartered. MacArthur felt that if the Formosa-based Nationalists attacked the mainland Communists that would leave the North Koreans to their own devices, as the Chinese wouldn't be able to carry on two wars at once.
    That type of thinking did not sit well with Bradley, who said

MacArthur was trying to get the U.S. "in the wrong war, at the wrong place, at the wrong time, and with the wrong enemy." His statement provided more support for MacArthur's ouster.

A third allegation against MacArthur was that he wouldn't listen to anybody but his most trusted advisors, particularly his G-2 (military intelligence) head, Charles Willoughby, who were not always trusted by everyone else. Those feelings are summed up by historian Bruce Cumings:

"[MacArthur and Willoughby] trusted only themselves, and had an intuitive approach to intelligence that mingled the hard facts of enemy capability with hunches about the enemy's presumed ethnic and racial qualities ("Chinamen can't fight"). This combined with MacArthur's "personal infallibility theory of intelligence," in which he "created his own intelligence organization, interpreted its results and acted upon his own analysis" (Cumings 25-26).

Comments and criticisms by MacArthur's contemporaries piled up. Combined with his strong views about using atomic bombs in Korea and expanding the war to China and Russia, which did not align with Truman's own, they were helping the president build a case for firing him. MacArthur was still advocating the use of nuclear weapons after he was relieved of duty and returned home.

In the spring of 1951 two U.S. senators, Richard Russell (D-GA) and Tom Connally (D-TX), conducted joint hearings to investigate Truman's handling of the MacArthur matter and to try to understand MacArthur's side of the argument. MacArthur did not help himself when he told the investigators that part of his problem was caused by the politicians in Washington who had introduced "a new concept into military operations—the concept of appeasement." And, he continued to advocate nuclear war during the hearings.

The general told the investigators that if the UN won the war

"...you will put off the possibility of a third world war." They asked him about the risk of a nuclear war with Russia. MacArthur told them that if such a scenario occurred, it would not be a good thing for the U.S. since, in his opinion, it was "rather inadequately prepared." Nevertheless, he intimated, the U.S. should not back away from such a confrontation.

Senator Theodore Green (D-RI) asked MacArthur what he thought would happen if the rest of the world shunned a nuclear war. We had better "go it alone," MacArthur replied. By the time the hearings concluded MacArthur's star had faded, in part because of his views on and advocacy of nuclear war if warranted.

Always lost in the shuffle has been the idea that numerous U.S. political and military leaders also espoused the use of nuclear weapons to the war. None of them were fired for their views, however.

Even after MacArthur faded away there was a long list of nuclear weapons adherents. MacArthur was the only one who paid the prices for suggesting their use. Perhaps it was the way he envisioned using them that appalled his critics.

Ike Meets "Mac" At John's House

MacArthur's actual plan for using nuclear weapons wasn't revealed until after he died. In a 9 April 1964 story published in the New York Times, MacArthur revealed that he "would have dropped 30 or so atomic bombs . . . strung across the neck of Manchuria...introduced half a million Chinese Nationalist troops at the Yalu...[and] spread behind us -- from the Sea of Japan to the Yellow Sea -- a belt of radioactive cobalt." That was a variation of what he had told President Eisenhower at a highly publicized 17 December 1952 meeting at John Dulles' house in New York. (MacArthur lived in New York City at the time as well.)

It was a major concession for Eisenhower to even sit down with MacArthur, who was considered to be the fifth most disliked person on Ike's list of disliked persons (Ike's Top 5...). (The other four, ranked in descending order were John F. Kennedy, British Field Marshal Bernard Montgomery, Harry S. Truman, and U.S. Senator Joseph McCarthy.) Eisenhower's reasons for not liking MacArthur were varied:

"He disliked MacArthur for his vanity, his penchant for theatrics, and for what Eisenhower perceived as "irrational" behavior.} He said, "Probably no one has had tougher fights with a senior than I had with MacArthur."

While Eisenhower served as Chief of Staff after World War II, MacArthur undermined his efforts to slow down mobilization and later to unify the armed services. Ike admitted, though, that MacArthur was smart, decisive, and a brilliant military mind. Working under him was frustrating, according to Ike, but also an invaluable learning experience for him.

Despite his feelings, Ike, who had served as MacArthur's assistant in Washington and his advisor in the Philippines in the 1930s, granted his former boss some time for a confab.

MacArthur told Eisenhower at the meeting that he had a plan to win the war. Bear in mind that MacArthur had been relieved of his command for 20 months at the time. But, technically he was still a general in the U.S. Army, so he felt comfortable offering a blueprint for victory. He suggested:

- A face-to-face meeting between Eisenhower and Stalin
- That the U.S. demand unity for Germany and Korea (Germany was divided into East and West at the time; Russia governed East Germany)
- That the U.S. and Russia mutually guarantee the neutrality of those two countries and Japan and Austria as well
- MacArthur had a few suggestions to implement if the Russians did not agree to those ideas:
- Remove enemy forces from above the 38th Parallel by

dropping atomic bombs on Communist military concentrations and installments in North Korea and military and industrial facilities in China
- Create fields of suitable radioactive materials, e.g. cobalt
- Close major enemy supply lines and disrupt communications from the Yalu River south
- Conduct amphibious landings on both coasts of Korea at the same time

Eisenhower listened to MacArthur's ideas and promised to consider them. That is as far as they went.

Make The Enemy Cobalt Blue

Later MacArthur explained that cobalt "has an active life of between 60 and 120 years. For at least 60 years there could have been no land invasion of Korea from the North." That was not 100% true. It's not like cobalt-induced radiation would have covered the entire area that he proposed bombing. According to scientific research, there are pockets in cobalt fields in which life can be sustained. But, if MacArthur had implemented his plan it would have put a serious crimp in the Communists' plans to occupy the southern part of Korea.

Cobalt bombs are also known as "salted" or "dirty" bombs. The cobalt itself, which is a hard, shiny, silver-white metal that is often mixed with other metals, could be used as a contaminant in an improvised nuclear device. Its presence would make the fallout more radioactive. The resulting radioactivity can last up to a century in length. The results of MacArthur's plan would have been devastating.

Why MacArthur chose cobalt as a weapon is a mystery. It is questionable whether he had any idea of how Cobalt-60 would affect the environment or troops who might be exposed to the radiation. The idea of using Cobalt as a weapon wasn't even introduced until five months before the Korean War began. Even then it was not a serious discussion.

Szilard Is Not Lizards Spelled Backwards, Sideways, Or Any Other Way

Physicist/molecular biologist Leo Szilard, who might be considered one of the fathers of the nuclear bomb, introduced the idea of a cobalt bomb on a February 1950 radio show. He did not have an actual bomb in mind. He simply used the idea of a cobalt weapon as a "doomsday bomb," one that could wipe out the world. Ironically, even though he conceived the idea of nuclear bombs, he was not a great fan of their use.

Prior to WWII, Szilard pushed for the idea of sending a confidential letter to President Franklin D. Roosevelt to introduce the feasibility of developing nuclear weapons. Albert Einstein endorsed his idea in August 1939. Szilard did more than talk about developing nuclear weapons. He and Enrico Fermi worked together at the University of Chicago to work on the project. They were a formidable team. Szilard and Fermi helped build the first "neutronic reactor," the device in which the first self-sustaining nuclear chain reaction was achieved. That was in 1942.

Naively, Szilard thought that if he and his colleagues developed nuclear weapons only one country could use them humanely. That was the ultimate oxymoron. As their work progressed and the idea that nuclear bombs were indeed a reality, the U.S. military started playing a leading role in the project.

U.S. Army General Leslie Groves, the military head of the project, assumed more and more influence. That displeased Szilard, who fought hard to curb Groves' power and dissuade the U.S. from using the atomic bombs ultimately produced.

Szilard was in the camp of people who believed that the threat of using nuclear bombs was enough to bring enemies to the peace talks table. He was dismayed when President Truman actually employed them in Japan. That accounts for Szilard's change in specialties after WWII from physicist to molecular biologist. It is ironic, then, that he should be the one to introduce the concept of a cobalt bomb, an idea which MacArthur picked up somewhere along the line during the Korean War.

There were no cobalt bombs available and knowledge about its properties and effects on troops was not highly developed at that point—not that the effects mattered. The U.S. military did not hesitate to put troops near nuclear bomb test or actual blast sites. (See ensuing discussions about troops in Nagasaki in September 1945, at Operation Crossroads in 1946, and at Frenchman Flats in Nevada in 1953.) The U.S. Air Force had looked into the possibility of acquiring cobalt bombs, but never followed up on the idea.

To experts' knowledge only one cobalt bomb was ever developed and tested. That wasn't until four years after the Korean War, and it wasn't a true cobalt bomb. On 14 September 1957 the British tested a bomb that used cobalt pellets as a radiochemical tracer for estimating yield. As the brief description explains:

Operation Antler

Test:	Round 1
Time:	0505 (GMT), 14:35 (local time); 1 Set 1957
Location:	Maralinga (Tadje)
Test Height and Type:	31 m aluminum tower
Yield:	1 kt

Test of Pixie, a lightweight small diameter implosion system with a plutonium core. This test later became notorious because of the experimental use of cobalt metal pellets as a test diagnostic or measuring yield (presumably by estimating the neutron flux from the degree of activation of the target pellet). Discovery of (mildly) radioactive cobalt pellets around the test site later gave rise to rumors that the British had been developing a "cobalt bomb" radiological weapon (http://nuclearweaponarchive.org/Uk/UKTesting.html).

Whatever results the British were looking for they didn't achieve. The test was deemed a failure. That ended any attempts to pursue the development of a cobalt bomb. It was too late for MacArthur, no matter what the tests proved.

Finally, in April 1951, Truman relieved MacArthur of his Korean command. That did not end the discussion about the use of nuclear weapons in Korea, though. President Eisenhower took up the argument where President Truman left off.

Chapter 8
Truman May Be Gone, But He Did Not Take The Nuclear Weapons With Him

"Japan learned from the bombings of Hiroshima and Nagasaki that the tragedy wrought by nuclear weapons must never be repeated and that humanity and nuclear weapons cannot coexist." Daisaku Ikeda

The U.S., encouraged by then presidential candidate Eisenhower, threatened the Communists with the use of nuclear weapons if they didn't at least discuss peace talks. That was a shrewd political move by Eisenhower, who knew that the war was not popular in the United States and Americans wanted it to end. He promised to go to Korea if elected and end the war. That was not something he really wanted to do, but a campaign promise is a campaign promise, and Truman backed him into a corner, so he had no choice but to go.

During the 1952 presidential race, Ike used the stalemate in Korea as a focal point in his bid to beat Truman. In a nationally televised 24 October speech in Detroit he labeled Korea as "the burial ground for 20,000 American dead." (The number at the end of the war exceeded 33,000+ Americans killed in action and 54,000+ killed from all causes in and outside Korea.) Furthermore, he promised, if elected, "I shall go to Korea."

In retrospect, there was nothing he could accomplish by going there. His WWII counterpart, Omar Bradley, admitted as much. "Ike was well informed on all aspects of the Korean War and the delicacy of the armistice negotiations," he said. "He knew very well that he could achieve nothing by going to Korea." Even Truman knew that!

The defeated Truman wrote in his diary, "I sincerely wish he didn't have to make the trip. It is an awful risk. If he should fail to come back I wonder what would happen. May God protect him."

Truman did not have to worry. Ike simply continued to put his faith in the threat of nuclear weapons, rather than the use.

Diplomacy Is Swell: Nuclear Weapons Are Better

Ike was a firm believer in nuclear weapons as a cornerstone of his foreign policy approach, even though he was not keen on actually employing them. He made his views on that clear from the outset of his presidency:

"The main elements of the New Look were: (1) maintaining the vitality of the U.S. economy while still building sufficient strength to prosecute the Cold War; (2) relying on nuclear weapons to deter Communist aggression or, if necessary, to fight a war; (3) using the Central Intelligence Agency (CIA) to carry out secret or covert actions against governments or leaders "directly or indirectly responsive to Soviet control"; and (4) strengthening allies and winning the friendship of nonaligned governments" (American President).

Eisenhower was not averse to spending money to support his advocacy for nuclear weapons. "His defense policies, which aimed at providing 'more bang for the buck,' cut spending on conventional forces while increasing the budget for the Air Force and for nuclear weapons" (American President). He made that clear to everybody—including the Communists in Korea. In his aforementioned "Cross of Iron" speech he also stated that nuclear weapons were not the only tool in the "peace bag."

"The way chosen by the United States was plainly marked by a few clear precepts, which govern its conduct in world affairs.

First: No people on earth can be held, as a people, to be an enemy, for all humanity shares the common hunger for peace and fellowship and justice.

Second: No nation's security and well-being can be lastingly achieved in isolation but only in effective cooperation with fellow nations.

Third: Any nation's right to a form of government and an economic system of its own choosing is inalienable.

Fourth: Any nation's attempt to dictate to other nations their form of government is indefensible.

And fifth: A nation's hope of lasting peace cannot be firmly based upon any race in armaments but rather upon just relations and honest understanding with all other nations.

In the light of these principles the citizens of the United States defined the way they proposed to follow, through the aftermath of war, toward true peace."

Ike's ideas of peace were still transcended by war, which was part of his inheritance from Truman.

In that same speech he blamed Russia for the war in Korea and noted how bad atomic war really was:

"The free nations, most solemnly and repeatedly, have assured the Soviet Union that their firm association has never had any aggressive purpose whatsoever. Soviet leaders, however, have seemed to persuade themselves, or tried to persuade their people, otherwise.

And so it has come to pass that the Soviet Union itself has shared and suffered the very fears it has fostered in the rest of the world.

This has been the way of life forged by 8 years [between WWII and the Korean War] of fear and force.

What can the world, or any nation in it, hope for if no turning is found on this dread road?

The worst to be feared and the best to be expected can be simply stated.

The worst is atomic war. (Italics added by author.)

The best would be this: a life of perpetual fear and tension; a burden of arms draining the wealth and the labor of all peoples; a wasting of strength that defies the American system or the Soviet system or any system to achieve true abundance and happiness for the peoples of this earth."

Talk Is Cheap, Bombs Are Expensive—But We'll Use Them Anyway

Eisenhower won the 1952 presidential election. He went to Korea as promised, and he continued to threaten the Communists with nuclear weapons if they did not engage in serious peace talks, which was a continuation of his predecessor Harry S. Truman's policies. He did not come back with any solutions in hand, nor did he formulate any in the aftermath.

Like his predecessor, Eisenhower was theoretically willing to employ nuclear weapons, or at least talk about using them despite the JCS's reservations. He said "…it was clear that we would have to use atomic weapons. This necessity was suggested to me by General MacArthur while I, as President-elect, was still living in New York. The Joint Chiefs of Staff were pessimistic about the feasibility of using tactical atomic weapons on front-line positions, in view of the extensive underground fortifications which the Chinese Communists had been able to construct; but such weapons would obviously be effective for strategic targets in North Korea, Manchuria, and on the Chinese coast."

He continued. "If we decided upon a major, new type of offensive, the present policies would have to be changed and the new ones agreed to by our allies. Foremost would be the proposed use of atomic weapons. In this respect American views have always differed somewhat from those of some of our allies. For the British, for example, the use of atomic weapons in war at that time would have been a decision of the gravest kind. My feeling was then, and still remains, that it would be impossible for the United States to maintain the military commitments which it now sustains around the world (without turning into a garrison state) did we not possess atomic weapons and the will to use them when necessary."

Ike realized that his decision to employ nuclear weapons would not sit well with his allies. "But an American decision to use them at that time would have created strong disrupting feelings between ourselves and our allies," he concluded. "However, if an all-out offensive should be highly successful, I felt that the rifts so caused could, in time, be repaired."

Finally, he observed, "...there were other problems, not the least of which would be the possibility of the Soviet Union entering the war.** In nuclear warfare the Chinese Communists would have been able to do little. But we knew that the Soviets had atomic weapons in quantity and estimated that they would soon explode a hydrogen device. Of all the Asian targets which might be subjected to Soviet bombing, I was most concerned about the unprotected cities of Japan" (Eisenhower 180).

** It is interesting to note that Ike mentioned the "possibility of the Soviet Union entering the war" when it was an open secret that they were already involved.

In the long run, Ike's threats did no more than Truman's or anybody else's to stop the fighting, which had reverted to a WWI trench warfare style. He had weapons other than atomic bombs in his bag of tricks to use as threats, however.

Ike Proposes "Annie"

One of the weapons Ike threatened to use was the new M65 cannon, which had the capability to fire nuclear or conventional shells. It was also known as a 280mm cannon, since its bore (the interior diameter of its gun barrel) was 280 millimeters wide. (We will call it "Annie" throughout the rest of this book.)

Some American troops were very impressed with its use, and

ultimately lauded its role in encouraging the Communists to sign a cease fire which took effect at 10 p.m. on 27 July 1953. Others, such as Frank Imparato, said they had no knowledge of its existence.

"I was in Korea during most of 1952 as an Infantry Advisor assigned to I ROK Corps as a Sergeant Major, and I never heard anything about 'Atomic Annie,' the 280 MM cannon," he revealed. (More of sightings vs. non-sightings later.)

Those who thought they had were convinced it had been fired at Communist troops a couple times just to demonstrate to them how horrific a weapon it could be. More than likely, they confused it with the 240mm cannon, which was deployed to Korea.

There is no doubt that the 240mm cannon was used in Korea, but not until early 1953, according to Donald L. Parrott. He recalled that it was fired in the last four months of the war by the 213th and 159th Field Artillery (F.A.) Battalions, U.S. Army. Based on his account, it is not likely that the U.S. Army would introduce both the 240mm and the 280mm around the same time in Korea. But, as the old saying goes, "military intelligence is an oxymoron."

"I served with the 213th and we were located in the Chorwon Valley in the Central Sector," Parrott noted. "In January 1953 our battalion had 155mm howitzers, and then to our surprise we received the (240) cannons in April 1953.

"Battalion had three firing batteries and I was with 'Charlie' Battery. The 159th F.A. Bn. covered the western half of the Korean Peninsula with their 6 (240) cannons. The 213th F.A. Bn. was to cover the eastern half of the Peninsula. At one time we fired both the 240 cannons and 155mm howitzers in May 1953. We eventually turned our 155mm howitzers over to the South Korean Army.

"The first week of June 1953 the 213th started to move to the east. 'Able' Battery remained in the Chorwon Valley. 'Baker'

Battery moved with their two cannons over to the Kumwha area, and my 'Charlie' Battery moved the farthest to the east--115 miles--to locate northeast of Yanggu and southwest of the 'Punch Bowl.'"

"It was a challenging move for us with 'Charlie' Battery. It included a ten-day stopover in the Hwachon Reservoir area. We fired the 240 cannon up to the cease fire on July 27, 1953. It was a potent weapon.

"The 240 cannon had a 9-inch diameter tube and could fire a 360-pound shell up to 14.3 miles. The tube itself was 22 feet long," he concluded.

The Case of the Disappearing Hill

Another veteran, Bob Barfield, a member of the U.S. Army's 187th Regimental Combat Team, noted the 240's impressive size too. "I couldn't believe the size of it and didn't know they had such a weapon in Korea," he said. "I looked it up on the web. It was used extensively in WWII, and twelve of them were taken out of mothballs and used in Korea."
(See https://en.wikipedia.org/wiki/240_mm_howitzer_M1)

Barfield revealed that the 240mm gun was first fired in Korea on May1, 1953, just before "our" battle for Boomerang. As he explained, "The site said that it was first fired as a ceremonial shot (whatever that meant) at a hill called the 'Donut.' The shot struck an ammo dump on top of the hill and set off a chain reaction that blew off the top of the hill, causing spectacular fireworks!"

The idea created speculation in Barfield's mind, much as it did in the minds of many Korean War infantry veterans. "Can you imagine if we could have used about 3or 4 of them against the Chinese?" he mused. "I'll bet they and the North Koreans would have thought, 'Damn, what the hell do those Yankees have now?'"

Barfield was awed by the gun's dimensions. "The cannon barrel was 27'6" long and could fire a shell that weighed 360 pounds a distance of 14.3 miles! It had a crew of 14 men. How in hell they ever got it up those hills is a mystery to me. That gun was a 'mountain buster.' Wow!"

Size Matters

The size alone of the 240mm cannon suggests that it could easily have been mistaken for a 280mm. The one major difference was the fact that the 280mm cannon could fire nuclear shells. The 240mm cannon could not. Whether the 280mm did fire atomic shells in Korea is a matter of speculation--and Korean War veterans are still speculating about whether it did.

Their convictions aside, there is no real evidence that there was ever a 280mm cannon in Korea between 1950 and 1953, let alone one that was fired. Then again, there is no evidence that it wasn't. The myths about "Atomic Annie" were bigger than its millimeters. The 280mm cannon may be the only artillery piece that ever helped win a war without actually participating in it, as the following pages suggest.

Chapter 9
The Birth And Development Of "Atomic Annie"

"One has to look out for engineers - they begin with sewing machines and end up with the atomic bomb."
Marcel Pagnol

If there had been a birth announcement for "Atomic Annie" it might have read like this:

The United States Army is pleased to announce the birth of "Atomic Annie." She was conceived at the Pentagon in Washington DC and born at Picatinny Arsenal in New Jersey. The father is Robert Schwartz, an engineer. The mother is "Anzio Annie," of Germany. "Atomic Annie" will most likely reside in Europe and possibly Korea when she is old enough to go out on her own.

Okay, that pseudo announcements needs a bit of elaboration.

Atomic Bullets And Atomic Cannons

The U.S. Army saw a need after WWII for an artillery piece capable of delivering nuclear shells on future enemies, whoever they might be. The primary enemy the Army had in mind was Russia, since it had engaged the U.S. in what came to be known as the Cold War following the end of WWII. So, in 1949 the U.S. Army assigned the task of developing a small tactical nuclear weapon to an "all-star" team at the Pentagon. It already had an "atomic bullet." A cannon would just be bigger weapon.

Actually, the "atomic bullet" was a figment of Army trainers' imaginations. Robert Jenkins recalls learning about it while undergoing basic training at Ft. Ord, CA in the spring of 1952 with the 6th Infantry Division, 63 Infantry Regiment, Company G.

"One day, while on a field of instruction, the cadre carried out what they said was an 'atomic bullet.' They carefully took it out of a metal box and loaded it into an M1 rifle and fired it down field. There was a big explosion at the point of impact...we were all amazed.

"It was a 'light hearted' moment when they told us afterward that the explosion was actually the result of explosives planted earlier."

The story about the "atomic bullet" may have been humorous, but the development of "Atomic Annie" was not. The Army was dead serious about acquiring a nuclear-capable artillery piece, and the sooner the better.

From The Pentagon To Picatinny

Two of the key development players were ordnance engineer Robert M. Schwartz, who drew up the preliminary designs for "Annie," and Samuel Feltman, whose job was to sell the final design to the Pentagon once Schwartz finished his work. Both were well suited for their roles.

Schwartz, a graduate of City College of New York and a former U.S. Navy radar project officer, joined Picatinny following WWII. His instructions were to work on a nuclear artillery shell, and to do so in utmost privacy. Then he could develop a cannon from which to fire it. Schwartz was never one to turn down a challenge. He was so eager to accept this one that he was willing to work in a locked room at the Pentagon.

His first step was to expand the size of the 240mm artillery shell to a 280mm shell. And, the new shell had to be a lot tougher than the 240mm shell, since it would travel at 6,000 rpms (revolutions per minute). That would only be possible if the

280mm shell was 4,000 times stronger than the casing of an atomic bomb. No problem.

Schwartz finished his preliminary sketches in only fifteen days. He returned immediately to Picatinny and worked in another locked room to fine tune his original design. By 1950 he had what he wanted and he assembled a team to pursue the development project under his lead.

Feltman had played a critical part in the development of ENIAC (Electrical Numerical Integrator and Calculator), the first electronic computer, at the University of Pennsylvania in 1946, which was funded by the U.S. Army. He secured the funding for the project and acted as the liaison between the Army's Ballistics Division and the key developers, John Mauchly and J. Presper Eckert.

Picking Up The Pace At Picatinny

The work at the Pentagon was transferred eventually to the Picatinny Arsenal in Dover, N.J., the headquarters of the United States Army Armament Research, Development and Engineering Center. Schwartz' initial idea was to cross the largest artillery piece at the Army's disposal at the time, a 240mm cannon, with the German K5 railroad gun known as "Anzio Annie" that had been used with great effectiveness against allied troops in Europe, most notably during their landing at Anzio, Italy in January 1944.

Pentagon officials liked what they saw in Schwartz's designs and gave the project a go-ahead. Thus began a three-year developmental project—at approximately the same time the U.S. became engaged in war once again in a place called Korea, which many Americans could not find on a map.

When Schwartz began his work, the U.S. president was Harry S. Truman, who had approved the use of nuclear weapons in Japan in 1945 in an effort to bring WWII to an end. As a result, Truman was not a great fan of using nuclear weapons, but he was certainly willing to threaten to use them, regardless of what newspapers reported.

Despite allegations to the contrary, Truman suggested that he did not really use them as a threat. He noted that news reports were responsible for that idea. He explained in a 1950 meeting with Clement Atlee, the Prime Minister of England, that "In spite of this assurance that the use of the atomic bomb was still subject to my approval and that I had not given such approval, news reports persisted that I had threatened to use the A-bomb in Korea" (Harry Truman 396).

Regardless of whom he believed was responsible, the threat was made clear to the Communists. But, even when told that nuclear weapons might help shorten the war, he still refused to approve their use.

Dean Rusk, Assistant Secretary of State for Far Eastern Affairs early in the Korean War, recalled that when it came to using nuclear weapons, Truman said no. Rusk cited one specific incident: "Only once do I recall serious discussion about using nuclear weapons: when we thought about bombing a large dam on the Yalu River. General Hoyt Vandenberg, Air Force chief of staff, personally had gone to Korea, flown a plane over the dam, and dropped our biggest conventional bomb on it. It made only a little scar on the dam's surface. He returned to Washington and told us that we could knock the dam out only with nuclear weapons. Truman refused" (McCullough 833).

Eisenhower was a bit more amenable to their use, even though he felt inwardly that to employ them would be bad public policy. His ascendancy to the presidency, the ongoing Korean War (which Truman insisted on calling a conflict or police action), and the development of "Atomic Annie" continued along parallel lines between 1950 and 1953.

Atomic Cannons and Nuclear Weapons

How Big A Truck Do You Need?

By 1951, Schwartz had a prototype nuclear cannon ready to test. It was a bit large: 84 feet long, 16.1 feet wide, and 12.2 feet high. The piece, including the gun and carriage, weighed 83.3 tons (166,000 pounds). The barrel length was 38.5 feet, slightly one-half the length of "Anzio Annie's." The 280mm bore (the mouth of the barrel) was 11 inches wide. "Atomic Annie" could fire conventional and nuclear rounds up to an estimated maximum range of 18 miles. That was one super cannon. But, like most artillery, it could not remain stationary, lest it attract the enemy's attention and subsequent attacks or the battleground moved.

"Atomic Annie" was the largest road-transported mobile artillery piece the U.S. Army ever employed. Building a "super gun" was one thing; transporting it was another. Artillery had to be mobile. The Germans had faced the same problem with their K5s. They solved it by placing them on railroad cars. The solution was feasible, but it placed a limitation on the cannons' use—especially since their gun barrels alone were 71 feet long. The large German guns could only fire at targets close to the tracks. The U.S. Army wanted to avoid that limitation.

Engineers devised a transport vehicle for "Annie" that included two separate tractors, each of which featured independent steering. It wasn't pretty, but it served its purpose. The vehicle could reach speeds of 35 miles per hour and make right angle turns on virtually any type of road 28 feet or wider, paved or unpaved. The crew could detach the cannon from the vehicle in fifteen minutes, fire, reattach it, and be back on the road in another fifteen minutes. That was a decided improvement on the Germans' rail-bound cannon, and gave "Annie" an expanded range over its predecessor.

The only thing left to do was test the new cannon—and maybe get it to Korea before President Eisenhower arrived there.

That did not happen, but he got to see one anyway. There was a prototype included in his inaugural parade on 20 January 1953, which Stan Britton recalls seeing: "There was one at 'Ike's' parade. I attended President Eisenhower's inaugural parade in Washington DC. There was an atomic cannon in it. It was huge, and had a truck at each end." But, he averred, "I never saw one in Korea, where I spent one year with the 78th AAA Gun Bn. 90mm."

Marvin H. Schafer did see one in Korea—somewhere—or at least he thought he did. "We saw an atomic cannon somewhere in 1953. Someone took photos of it because of its unusual nature, because it was called an atomic cannon, and because it had to be loaded with a crane. I do not know on what army base the photo was taken or who the soldier is."

Testing, Testing, 1, 2, 3...

By May 1953 the Army was ready to conduct its first test of "Annie." At 8:30 a.m. on 25 May 1953, the Artillery Test Unit from the Artillery Center, Fort Sill, Oklahoma, tested the cannon at Frenchman Flat at the Nevada Test Site, which had been established on 11 January 1951 for testing nuclear devices. The site covered about 1,360 square miles comprising desert and mountainous terrain. The event was well attended. Among the 500 guests were Chairman of the Joint Chiefs of Staff Admiral Arthur W. Radford and Secretary of Defense Charles Erwin Wilson. They were not disappointed.

The crew successfully fired a 15 kiloton shell with a W9 warhead at a range of seven miles. (A kiloton is a measurement of the yield of a nuclear weapon equivalent to the explosion of 1,000 tons of TNT.) The large W9 was 11 inches (280 mm) in diameter, 55 inches (138 cm) long, and weighed 850 pounds (364 kg). It detonated successfully—and went down in history as the only nuclear shell ever fired from a cannon (at least officially). The U.S. did not waste any time letting the world know that it had an atomic cannon.

Secretary of State John Foster Dulles happened to be on a trip to the Middle East and Asia in May 1953. He was visiting Prime Minister Nehru in India when he revealed the news about the cannon. He told Nehru in comments intended for Chinese Premier Chou En-Lai that the U.S. would not only start bombing Manchuria if the Communists didn't start bargaining for a peace settlement, but it might start using the new cannons in Korea. Thus, "Annie" became a psychological weapon practically at birth. The Chinese were not duly impressed, apparently. They stepped up their efforts in Korea despite the warning—if Nehru ever delivered it.

Let's Give It A Shot

The first test of the atomic cannon was an event to remember, especially for the soldiers assigned to "attack" the blast site.

In addition to the guests, there were 2,500 soldiers in positions to witness the event. Some of them were huddled in trenches less than two miles away from the blast as part of a tactical exercise. A few minutes after the explosion occurred, they charged towards the test site—with the mushroom cloud still hanging overhead.

Units became disoriented in the swirling dust churned up by the winds created by the tremendous pressure of the nuclear explosion. As veteran William Russell, a former U.S. Army Combat Correspondent, exclaimed, "One wonders how many of those soldiers survived or how their lives were probably shortened by this foolhardy exercise." Actually, there was precedent for having troops up close and personal at the site of an atomic blast.

Damn The Bomb: Send In The Troops Anyway

The developers of the atomic bomb had not done much

testing to discern the effects of radiation on troops or equipment before dropping the atomic bombs in Japan. Thus, just a few weeks after the air crews delivered "Little Boy" and "Fat Man," infantry troops were deployed to Hiroshima and Nagasaki for occupation duty.

The 186th Infantry Regiment of the 41st Division, X Corps of the Sixth Army deployed to Hiroshima about sixty days after the bomb was dropped there. It was replaced eventually by the 34th Infantry Regiment of the 24th Division. Marines arrived in Nagasaki only 45 days after the bomb fell there. There were some precautions taken before the troops entered the cities.

Groups of American scientists from the Manhattan Engineer District arrived in each city three days before the troops to perform radiological surveys. In at least one case it was too little too late. The repatriation of former prisoners of war (POWs) through Nagasaki began before the survey was completed and the city was occupied.

Nagasaki was used to repatriate former POWs because the waterfront was sufficiently far enough away from the hypocenter (the spot on the ground directly under the detonation, i.e., ground zero) to have escaped most of the destructive effects of the bomb and to have been free of radioactivity. Over 9,000 allied POWs, including 2,300 Americans, were processed at Nagasaki from September 11-23, 1945 (DRTA 2).

The total number of troops occupying Hiroshima was formerly estimated to be about 40,000, and approximately 27,000 troops occupied Nagasaki. About 12,000 troops occupied outlying areas within 10 miles of either city through July 1, 1946. An additional 118,000 servicemen or more had passed through these areas by July 1, 1946. These transient personnel included POWs, troops disembarked for elsewhere in Japan, and crews of ships docked nearby. In total, the Nuclear Test Personnel Review Program identified over 230,000 veterans who participated in the occupation of Hiroshima and/or Nagasaki (DTRA 1).

And, it wasn't until well after the war that tests were conducted to find out how radiation affected ships and other assets. That occurred when the U.S. Navy conducted Operation Crossroads, which turned out to be a fiasco.

Chapter 10
Marines Learn That Radioactive Doesn't Mean Changing The Radio Station

"Based on radiation surveys performed by American scientists from the Manhattan Engineer District prior to the arrival of the occupation forces, the greatly-decayed residual radioactivity levels in and around Hiroshima and Nagasaki at the time the occupation forces arrived were such that military activities could proceed as planned, unimpeded by radiological considerations." Department of Veterans Affairs (VA), Nuclear Test Personnel Review (NTPR) Program

One of the proposed uses of nuclear bombs in Korea was as anti-personnel weapons. Because of the "trench warfare" nature of the war, however, there was some questions as to whose personnel would suffer the most. Atomic bombs dropped on Chinese and North Koreans in all probability would have impacted UN forces as well.

Then there were the effects of radiation on everyone within an atomic blast area to consider. Anecdotal evidence suggested that troops could function normally in areas devastated by atomic weapons once the residual radioactivity levels subsided. The fact that many of them had survived their postings in Japan with no apparent ill effects after the atomic bombs fell seemed to be adequate proof that troops would be able to survive in Korea as well.

U.S. military and political leaders were not reluctant to place

American troops in areas nearby atomic blast sites in the early days of atomic weaponry. They did so in Japan shortly after the bombs were dropped on Hiroshima and Nagasaki and again during subsequent tests of more powerful nuclear and hydrogen weapons, e.g. at Frenchman Flats and Operation Crossroads. The troops so disposed were in effect test subjects to measure the impact of nuclear weapons on human beings. Whether that would have been the case in Korea is a moot point, since the bombs were never used. But....

Would the UN have warned the troops that atomic weapons were about to be used, or wait until after they were? Would they have issued special equipment to prevent radiation poisoning or other deleterious results? Would they have moved the troops away from their positions until the bombs were dropped, thereby running the risks of alerting the enemy that something different was about to happen?

There is no evidence in Korean War veterans' stories that any of these questions were considered by their leaders when they mulled the use of nuclear weapons in Korea. They did not discuss special training or equipment. Mostly their references to nuclear weapons were about rumors or alleged sightings of mysterious cannons that could fire atomic shells. No doubt those rumors reached the enemy. That was the goal of psychological warfare and the best use of nuclear weapons in Korea.

Send In The Marines

U.S. Marines from the Second Division arrived in Nagasaki in September 1945, almost before the dust had a chance to settle. Among them were the 2nd, 6th, and 8th Regimental Combat Teams (RCTs), artillerymen from the 10th Marine Regiment, members from a Headquarters Battalion, service troops, an engineer group, a tank battalion, an observation squadron, and representatives from smaller units. Granted, they were not nearby when the bomb was dropped, as the troops at Frenchman Flats were, but they ran a risk from the after-effects of radiation just the same.

The Marines, like their counterparts in the other services, entered the radioactive zones of Japan without questioning why. That was their duty. A few of the Marines described their experiences in their memoirs. Roland Jennings starts the story.

"For all practical purposes, the conflict in the Pacific Theater was over when the 'big one' was dropped on Nagasaki. We were at Saipan preparing for the November invasion of the Japanese mainland. On that decisive day, 6 August 1945, our outfit, Company C, 1st Battalion, 10th Regiment, was gathered in the mess hall attending a briefing on the island of Agrihan. The 1st Battalion was scheduled to participate in a live fire ARSOP (Artillery Reconnaissance, Selection and Occupation of Position) on this uninhabited island, allowing us to register our artillery on land-based targets, not into the sea, as was the norm on Saipan.

"We were knee deep into the briefing when a young, exuberant trooper from 'A' Battery raced up to the screened window of the mess hall, and blurted out something about the war being over. Suddenly, the room became very quiet. A stunned silence fell over the group before pandemonium set in.

"The officer presenting the briefing hung in there for a while, but any semblance of order had all but vanished. Elation abounded for the moment. Thirty days later, the entire Second Division was aboard ship en route to Nagasaki, ready to assume occupation duties on Japan's southernmost island, Kyushu. I've often wondered who was left behind to strike the tents etc. Perhaps it was elements of the Quartermaster Corps.

"We, the 10th Marines, left our campsite intact when we traversed the island and boarded ships. Picking up after an estimated 20,000 troops would have been a time consuming task. In fact, the Division lingered in Saipan's harbor for a few days, enjoying shore leave at the Navy's recreation area, located on the beach near Garapan, before weighing anchor.

"I had been overseas fifteen months, with two campaigns under my belt, when the atom bombs brought the war in the Pacific to a close. Second Marine Division personnel were about to assume the role of ambassadors per se, instead of military conquerors. As for me, I looked forward to seeing the nation that had challenged the USA four years ago by staging the infamous attack on Pearl Harbor.

"My outfit, the 10th Marines, landed at Nagasaki harbor on 23 September 1945. After a one-week hiatus at Isahaya, they were reassigned to the port city of Nagasaki, continuing to perform as military police. One of my first duty assignments was guarding the English embassy, located a few blocks south of the downtown area.

"The two-story English and American embassies stood side by side like silent sentinels. They were in a sad state of disrepair. Utilities such as electricity and water were not available within the structures. Our only source of lighting was our standard issue, two-cell flashlights.

"In peace time the embassies afforded a degree of comfort for the diplomats and their families. One of the more desirable amenities was a built-in fireplace. In contrast, the Japanese used a large stone urn filled with charcoal briquettes for their source of heat.

"Later in the evening, my buddy and I attempted to start a fire in the American embassy's fireplace by igniting the open end of an unwieldy log found on the premises that we had wrestled over to the hearth. Looking back, this was really a foolhardy idea. We very well could have duplicated another feat that was attributed to Mrs. O'Leary's cow in 1871, 'The Great Chicago Fire.' I don't believe the local authorities had much in the way of firefighting equipment.

"After our futile attempt failed, I turned my attention to exploring the interior of the English embassy, my assigned post. As I inched my way up the dark staircase, careful to avoid the debris on the steps, every shadow seemed to come alive. When I reached the landing, I peered into every nook and cranny, where I saw the terrazzo tiled bath, the bedrooms, the closets...all the while hoping I was alone in my endeavor. What I didn't need to find at this point was a pair of beady eyes staring back at me. The squalid conditions in and around the embassy would have made a perfect environment for the large rodents seeking a quiet haven.

"My flashlight was beginning to grow dim, so I beat a hasty retreat down the stairs and stayed within earshot of my buddy next door. We didn't discuss it, but he didn't appear to be too comfortable with his situation either. Finally, our shift drew to a close, allowing us to return to the safe confines of our barracks located across the bay. Chalk up another successful day for these two Marines. We were survivors.

"For the life of me, as a 19-year-old Marine, I couldn't imagine why we were assigned to guard a derelict sitting in the middle of Nagasaki, a city that had been struck by an atom bomb only a few weeks earlier. I suppose we treated this like other questionable assignments 'while on loan to the Corps.' But, it's not for us to wonder why, but for us to do or die.

"Looking back at the war years as a septuagenarian, I have a slightly different perspective on things and events. Guarding the embassies wasn't really the question. It may have been for a continued show of force and regaining respect by reclaiming real estate allotted to the USA and its allies by the Imperial Japanese government, a reciprocal action practiced during peace time."

Raymond Bistline had a similar story to tell about his memories of Nagasaki in 1945.

"I was a Corpsman in Company B - Med, Second Marine Division in early September, 1945, stationed on the island of

Saipan. We received word that Japan had surrendered and suddenly we became stevedores, loading our equipment onto the transport, the USS Melette (APA-156). When completed, our ship joined the convoy of 20 transports and 6 destroyers heading for Nagasaki, where the second atomic bomb had been dropped on August 8th. It took about ten days.

"It was early morning, 23 September 1945 when we entered Nagasaki harbor. As the fog lifted, we saw to our surprise a small Christian Church. Looking around us, we saw the tops of more than a dozen sunken Japanese ships in the harbor area. Over on the far left, we saw a small task force, consisting of a hospital ship, a small aircraft carrier, two cruisers, and two destroyers, busy taking on liberated POWs. We returned to the hold and began unloading food and other supplies for Marines who landed the first day.

"We continued this on the second day, when suddenly we heard the call: 'Grab your gear. We are going ashore in fifteen minutes.' I picked up my gear and ran out on deck. I looked over the side and saw LCVPs (Landing Craft, Vehicle, Personnel) circling the ship. When it came my turn, I climbed down the net and dropped into the boat. It soon joined the others to form a wave.

"We proceeded up river about five miles, past the large Mitsubishi Shipyards on the left, where the large Japanese battleships and aircraft carriers were built. We landed on a dock on the right. We climbed out with our gear and waited an hour for our truck.

"We slowly drove through the streets, which were crowded with people and lots of little children. They shouted 'Ohaya' (good morning). For four years we were taught to hate the Japanese. It took us ten minutes to begin to love them.

"We reached into our packs and threw candy, chewing gum, and cigarettes at the spectators. We drove about ten miles southeast to the former Japanese camp Camp Kamigi and unloaded our gear.

"Five days later, Phil Bard, Sidney Berkowitz, Gene Black,

and I were ordered to the medical supply tent. We loaded our truck with medical supplies for the atomic bomb victims so high that the four of us had to lie on top and pass the electric wires over the truck.

"We tried to help the Japanese in every way we could. We traveled freely through the undamaged part of Nagasaki and even went to a movie. But we were forbidden to enter the atomic bomb area. To see this devastated area, we had to climb a mountain and look down.

"We saw an area as large as West Philadelphia, completely destroyed with a large wreath in the hills, created by forest fires. I cried when I saw this massive destruction and where over 100,000 Japanese were killed. Every day we drank water so radioactive that its activity could not be measured on a Geiger counter.

"After a six-week period we moved to southeast Kyushu to Miyakonojo, then to Oita and Beppu, and finally Sasebo. It was on 20 June 1946 that we finally received orders to go home.

"Our hospital was located on a cliff overlooking the harbor. We saw two ships come into the docks. We were ordered to pick up our gear and go down the long S-shaped trail to the ships. The trail was lined with Japanese singing 'Auld Lange Syne.' We joined them in this song of farewell.

"We boarded the USS Rutland and began our 45-day journey home. It was a wonderful experience to see the Japanese become so completely a warm, kind, and considerate people."

Whether there was a valid reason for stationing troops in a city just leveled with an atomic bomb is a matter of speculation. Was it just to see how they reacted to the radioactive atmosphere? Or was it for public relations purposes or a cultural exchange experience? Those are the types of questions raised in the story presented by Rolland K. Hindsley, who served with G Company, 2nd Battalion, 8th Regiment, 2nd Marine Division.

"The 8th Marines were on Saipan when the two atomic

bombs were dropped on Japan. We were told we would be going to Japan as occupational troops.

"We sailed into Nagasaki harbor early in September 1945. After our ship, S.S. McIntire, was secured, our battalion went ashore with our rifles, ammo, and backpacks. After a while we asked where the rest rooms were. An interpreter told us where to go 'through some swinging doors, and you will find the 'Benjo.' That was our first experience with girls and men using the same bathroom.

"Later, we boarded trucks and we were taken a few blocks from the dock area. Two of us were ordered to stand on a corner till we were picked up later. After some time, 2 or 3 small boys were brave enough to approach us. They were about 6-7 years old.

"We gave them Hershey bars. I took off my helmet. These boys had never seen red hair and green eyes. My buddy took off his helmet and added to the display. He had blond hair and blue eyes. We couldn't understand what they were saying, but I'm sure they were surprised to see red and blond hair.

"Four days later we boarded a train for Kumamoto. We were stationed in a boys' military school, which presented good accommodations. After six months of occupational duty, I had enough points to go home. The 8th Marines had done their job, and I was a part of it."

These three Marines were just a few of the many who were stationed in a location affected by atomic weapons just a few weeks after the attacks. How many of them suffered ill effects from their experiences is unknown. Yet, U.S. political and military leaders were willing to take that same risk in Korea, after using much stronger weapons. Just how the troops would have reacted there was never learned, since the weapons were not used. Based on the three stories presented above, the troops no doubt would have undergone the experience as a part of their assignments.

Chapter 11
Operation Crossroads

"The bomb will not start a chain-reaction in the water converting it all to gas and letting the ships on all the oceans drop down to the bottom. It will not blow out the bottom of the sea and let all the water run down the hole. It will not destroy gravity. I am not an atomic playboy, as one of my critics labeled me, exploding these bombs to satisfy my personal whim." Admiral William H. P. Blandy

Less than a year after WWII ended the JCS decided to conduct more tests to collect data about the effects of radiation on weapons and ships and to assess new aircraft techniques for delivering nuclear weapons. JCS was seeking data in several areas. The first four involved scientific research, which accounted for the eclectic mix of researchers that included botanists, zoologists, geologists, and oceanographers from universities, oceanographic institutes, and government research bureaus:

- Yield (TNT equivalent)
- Blast (overpressure)
- Heat (thermal intensity)
- Radiation (both initial and residual effects)

The military folks wanted to learn about weapons effects, aircraft attack techniques, and naval ship design criteria. The implication was that the U.S. would be willing to use nuclear weapons in the future if necessary. Military leaders just wanted more information about how the people and equipment in the target area would be affected. The tests did not turn out quite the way the JCS expected. In fact, they were never completed.

Close, But No Cigar

The plan was to detonate three nuclear weapons at Bikini Atoll in the Marshall Islands in a July 1946 exercise dubbed Operation Crossroads, involving about 42,000 military and scientific personnel. The goals were to expand on the data collected after the one and only test of nuclear weapons, code named Trinity, and the explosions generated by "Fat Man" and "Little Boy." The information from those three events was limited, since the goal was to end the war, not gather scientific evidence.

The first segment, named Test Able, involved a bomb dropped from a B-29 at a low altitude. The target was a gaggle of 95 ships gathered inside Bikini Lagoon, all of which were destined for the scrap yards anyway. Many of them contained live ordnance and had animals aboard as experimental subjects. The second test, named Baker, involved the detonation of a bomb dangling 90 feet underwater from LSM-60. Third was Test Charlie, a planned deep water explosion.

Significantly, the three explosions would be only the fourth, fifth, and sixth times nuclear weapons were detonated. Two of the first three were the bombs dropped in Japan. The other was Trinity. That one, conducted by the U.S. Army, took place at the Alamogordo, New Mexico test site on July 16, 1945 as part of the Manhattan Project—only three weeks before the first nuclear bomb was dropped.

Things went amiss with Operation Crossroads right from the start. (Just to demonstrate how bad things got at times, one of the U.S. Navy ships bringing test samples for radiology and histological studies back in the United States ran aground off the coast of California, and most of the specimens were lost.)

Test Able took place on 1 July 1946. The B-29 flew in at the designated altitude level of 520 feet and the bomb fell. Unfortunately, it missed the target by 1,500 yards. As a result, only five ships sank. A few more sustained some damage. The blast did not provide the scientists or the military observers with much usable data. There was not enough residual radioactivity to measure. On to Plan B.

The Heck With The Fish: Let's Blow Up Some Ships

Test Baker was a little more helpful. The personnel gathered on 25 July to witness the explosion—and quite an explosion it was. It caused a column of water to rise about a mile high and a half-mile wide. Once the water came back down it initiated a heavy fog called a "Base Surge" that covered the entire fleet. That was what the scientists were looking for. The Base Surge contained the radiation equivalent to tons of radium. Its components comprised fission fragments (split uranium and plutonium atoms) and unfissioned plutonium.

In scientific terms, all of these components were highly radioactive with Gamma and Beta emitters and undetectable Alpha emissions. The profusion of radiation created a problem for the Lagoon Patrol and Initial Boarding Teams.

For five days the team members worked their way around the intense fields of radiation to beach sinking ships, retrieve test instruments, and evacuate animals from the target ships for further study. Navy personnel wanted to decontaminate affected ships and board them. Their efforts were thwarted, though, because the radiation levels did not abate. Undeterred, the researchers began to ignore safety procedures.

On 1 August 1946 specially assembled ammunition work teams started removing the live ordnance from the target ships, a process which continued into March of 1947. Vice Admiral W. H.

Blandy, commander of Joint Task Force-1, didn't stick around that long. On 10 August, he ended all decontamination attempts. Blandy and most of his staff left for home three days later. The ships were scattered in different directions.

There Is No Sense "Atoll" In Restricting The Radiation To Bikini

During August crews busied themselves towing target ships from Bikini Atoll to Kwajalein Atoll, where a personnel decontamination barge awaited. The plan was to reduce the radiation levels aboard the relocated ships at Kwajalein so researchers could do their work amid lower radiation levels.

Many of the radiation-laden ships were returned to the United States for further studies. They included eight major ships and two submarines. Twelve of the ships weren't affected by radiation and were returned to the United States. About sixty of the remaining ships were taken to different locations around the Pacific Ocean and sunk. Apparently everybody forgot about Test Charlie, which was canceled.

The entire operation was discontinued effective 1 November 1946. The JCS was given the discretion of resuming operations, but it did not.

So that's what the political and military leaders had to work with once the Korean War began. They had some basic information gleaned from the results of an operation that did not yield much helpful data as far as what would happen to troops on the ground should they choose to employ nuclear weapons. They had also learned something about the effects of radiation on ships, but ships would not be their primary targets.

Only four years elapsed between Operation Crossroads and the beginning of the Korean War. That was not enough time for mortality studies to be conducted regarding deaths due to radiation, and it is not likely that the military leaders would have paid attention anyway.

Just for the record, studies indicated that personnel exposed to radiation at Operation Crossroads did not exhibit mortality rates out of line with control groups:

"These findings do not support a hypothesis that exposure to ionizing radiation was the cause of increased mortality among CROSSROADS participants. Had radiation been a significant contributor to increased risk of mortality, we should have seen significantly increased mortality due to malignancies, particularly leukemia, in participants thought to have received higher radiation doses relative to participants with lower doses and to unexposed controls. The committee did not observe any such effects. They note however, that this study was neither intended nor designed to be an investigation of low-level radiations effects, per se, and it should not be interpreted as such (Mortality)."

In effect, Truman, Eisenhower, MacArthur et al had extremely limited information on the effects of radiation and nuclear explosions on various targets as the Korean War progressed. All they had was new and improved nuclear weapons and a nuclear cannon that had been tested once.

Chapter 12
At Least We Have A Nuclear Cannon

"A world without nuclear weapons would be less stable and more dangerous for all of us." Margaret Thatcher

Mission accomplished! The army had its nuclear cannon. The question was what to do with it. The first construction orders went out quickly.

Eventually, the Watervliet Arsenal in New York and the Watertown Arsenal in Massachusetts manufactured and assembled at least twenty M65 cannons. They were expensive, at $800,000 each. In the long run, the $16 million tab was not well spent. The cannons were deployed to Europe and Korea, but they were never used in combat, and went out of service in 1963.

The money was not completely wasted. "Atomic Annie" did become a psychological weapon in the U.S.-led efforts to end the fighting in Korea and facilitate a prisoner exchange. U.S. officials deliberately created an air of mystery about their "secret weapon" to keep the enemy off guard about its use, location, etc.--or even who owned it.

A POW's Knowledge

Chinese soldiers apparently knew about an atomic weapon in Korea, even if they didn't know which side was employing it. This became apparent in an interview with a Chinese prisoner of war named Ch'en Chin Sheng, who went to great lengths to get captured by UN troops in the later stages of the war. In fact, he wasn't taken prisoner until after the cease fire was signed on 27 July 1953, when he surrendered to British troops of the 1st

COMWELL Division.

If nothing else, Sheng was determined. He attributed his desire to escape in part to the psychological warfare conducted by UN forces, of which atomic weapons were a part, although he did not cite them specifically. His comments were included in a U.S. Amy, "Prisoner of War Preliminary Report," Headquarters I Corps, Office of the Assistant Chief of Staff, G2 document dated 01 August 1953.

"PW (Prisoner of War) thought of deserting long before he arrived in Korea because his father had been executed and property confiscated by the Communists. He claimed that UN leaflets reaffirmed his desire to surrender. CCF (Chinese Communist Forces) burned all leaflets and told the men that UN mistreated PW. PW did not believe this because he had seen Americans when he was in the CNA (Chinese National Army) and knew what to expect. PW believes that UN propaganda affected CCF personnel in that many are already dissatisfied with CCF life."

According to the interview conducted by U.S. Army Captain Chew-Mon Lee:

"Despite limited schooling, PW is intelligent and could read maps, when oriented. He willingly gave information known to him*, and had a good memory of dates and places, and was very cooperative. PW knew a cease-fire was near, but because he travelled (sic) three days and four nights to surrender, he did not know an armistice had actually been signed during this period. He experienced a strong desire to go to Formosa. Information given by PW is considered reliable. Not recommended for further interrogation."

* Unlike U.S. troops, who were instructed to reveal nothing more than their name, rank, and serial number to enemy interrogators after being captured, Communist troops in Korea were not restricted by any code of conduct from telling their captors what they knew. The theory was that they didn't know very

much, so whatever they revealed would not help their interrogators.

Sheng verified some of the information that UN forces already knew, e.g. that Russian troops had participated actively during the war. He told Captain Lee:

"[His] unit arrived in Antung...to supplement the Heavy weapons already with it. [He] also attended Korean orientation lectures and anti-air classes. When [he] went to the Yalu River to bathe, he saw Russian-manned 76mm AA batteries guarding both the Chinese and North Korean sides of a steel railroad bridge."

He saw more Russians later at Pyongyang, where he was assigned to dig trenches and repair roads.

One of the admissions he made might have startled UN leaders who were aiming for a peace pact of some kind, with or without atomic weapons. Sheng said that the Chinese were not really interested in peace:

"PW heard from unit officers that peace would not be a 'real' peace. PW stated the CCF never had peace in mind, and was told the CCF would build up the North Korean Army (NKA). If peace did not come they would openly assist in attacking the UN, and if peace did come, the CCF would wear NKA uniforms and attack as such. An offensive was to be launched in late October, just before winter, because 'Americans do not like to fight in ice and snow.' The unification of Korea by force was to be achieved by the end of 1953."

That last piece of information might have come as a surprise to the UN forces who were banking on atomic weapons as a tool in getting the Communists to the peace table. Granted, there was no proof that Cheng had reliable information about the Communists' intentions regarding peace, and the fighting had ended at this point. But, his statements raised a question about the role of atomic weapons in the eyes of the Communists in the peace-keeping process. And Cheng did allude to the presence of such weapons in

Korea, although he was a bit confused about who deployed them and what they were.

"PW heard rumors of a secret weapon called the K'A CH'IU SHA, which was used in August 52 during a 42-day battle (possibly Heartbreak Ridge). PW heard this was a Russian 'Atomic' weapon mounted on trucks." (Captain Lee noted that "This could possibly be the Katusha Rocket.") PW heard that in this battle the entire Turkish Brigade was annihilated."*

* That did not happen at Heartbreak Ridge or anyplace else in Korea where the Turks fought.

The suggestion of an "atomic" weapon at that time of the war was ironic. That was about the time that some of Sheng's U.S. counterparts allegedly saw "Atomic Annie" in Korea, so it is possible that soldiers on both sides were relying on rumors about atomic weapons being used on the battlefield. Whether they were there, and regardless of which side was using them, the rumors add to the mystery about "Atomic Annie's" role in the Korean War. That mystery persisted even among U.S. soldiers who were familiar with the cannon.

The Legend Lives

Arlen Pease, who had firsthand knowledge about "Annie," addressed the secret atmosphere around the cannon.

"I was assigned to the Department of Gunnery Enlisted Staff and Faculty Battery during 1955. The atomic cannon was part of the armament at that time, conventional shells (no atomic). The Army announced officially in 1952 that an artillery piece capable of firing an atomic shell had been developed, and the first atomic shell had been fired (I think at Yucca Flats, Nevada). But, there was no indication as to when it might go into service.

"The atomic shell was first fired as a test in 1953. I have no knowledge that it was ever sent to Korea…Since the first firing test was in 1953, and the truce was signed 27 July 1953, would it have been there during combat?" That is exactly the question that is hard to answer precisely.

Pease continued with a description of the gun that suggested its exact—or even approximate—dimensions were not known by soldiers in the field. That was a contributing factor in the mystery about whether "Annie" ever appeared in Korea.

"I think the weight of the cannon was in excess of 50 tons. The projectile weighed approximately 400 pounds; the range was 35,000+ yards. The cannon rested on two built-in platforms and was carried from place to place by two hoist-lift type motorized machines." He was close to an accurate description in some respects.

Finally, Pease said, "I spent 1956 in Korea as an Artillery Survey Party Chief in the 1st Field Artillery Observation Battalion, HQ Battery, I Corps. We interacted with all the UN artillery that was still in Korea, and I didn't see or hear of any equipment as mentioned. All firing with the cannon was done with conventional shells, except for the testing in 1953."

It was there one day, but not the next. In fact, it was sighted so often in so many different places people have to wonder if there weren't more than twenty of them. Arnold Van Deusen noted in the Sept/Oct 2011 edition of The Graybeards that he saw "Annie" in Maryland in 1952.

"I went to Aberdeen Proving Grounds for basic and tech training in April 1952," he wrote. "During one of our many-mile hikes we were ordered to break ranks and get off the road. Then, a huge two-part truck carrier went past hauling a massive cannon. The truck had two tractors with a drop-frame carrier in the middle. Drivers talked to each other by a radio intercom system. I didn't know what it was until months later, when we saw a photo in the Yank Magazine in Korea.

"I worked on the railroad in Korea, but I never saw the cannon over there. I don't doubt that it got there eventually, though."

He was not alone in that opinion. It didn't matter to Ike whether he had ever seen one or if "Annie" actually got to Korea. He used the M65s as part of his threat to employ nuclear efforts to end the war.

Again, whether any of them reached Korea during the actual fighting is debatable. But, as the comments made by Sheng, Pease, Van Deusen et al imply, some troops on both sides believed that "Annie," or some type of atomic weapon, did get there. In fact, some of them believed nuclear weapons were actually used in Korea.

The mere threat of their use was President Eisenhower's psychological weapon, even if the actual "Annies" were never employed. It was one of the tools he used to try to convince the Communists it would be to their benefit to bargain for peace, or at least a cease fire. If Sheng is to be believed, the threat did not work.

Chapter 13
Ike Visits Korea

"The use of the atomic bomb, with its indiscriminate killing of women and children, revolts my soul."
Herbert Hoover

Exactly what Ike did or did not do with respect to ending the Korean War is as mysterious as the presence or non-presence of "Atomic Annie" in Korea. Some of what he did is substantiated. Some is merely rumored. Historians and veterans differ on his role and his alleged threat to drop nuclear weapons on North Korea and/or China if the Communists did not get serious about peace negotiations. One thing is certain: "Atomic Annie" may not have gotten to Korea during the war. Ike—the president, not the cannon—did. .

The Drive To Korea Starts In Detroit

During Eisenhower's presidential campaign in 1952 he promised to go to Korea to end the war. In a 24 October 1952 speech in Detroit, just days before the presidential election, he blamed the Truman administration for turning a blind eye to the events that led up to the North Korean invasion of South Korea on 25 June 1950. And, he promised that ending the fighting in Korea would be the centerpiece of his foreign policy.

"Where will a new administration begin?" he asked rhetorically.

Ike answered his own question.

"It will begin with its President taking a simple, firm resolution. That resolution will be: to forego the diversions of politics and concentrate on the job of ending the Korean War—until that job is honorably done.

"That job requires a personal trip to Korea."

Then, he made his pledge.

"I shall make that trip. Only in that way could I learn how best to serve the American people in the cause of peace."

"I shall go to Korea," he reiterated.

"That is my second pledge to the American people."

He was as good as his word. He won the election, and he went to Korea shortly thereafter—even before he took office--on a trip that even Truman did not think was necessary. Nevertheless, Truman offered Ike the use of his plane, Independence, for the trip. Ike declined the offer.

The former president said in his diary, "I sincerely wish he didn't have to make the trip. It is an awful risk. If he should fail to come back I wonder what would happen. May God protect him" (McCullough 913). Truman's worst fears did not come to pass.

Amazingly, Americans did not even know Ike was in Korea, and the news about his trip was not made public until he was en route home. What Ike did there has long been a subject of speculation, although most historians have tracked his movements and meetings with preciseness.

It is known that he met with U.S. and UN military leaders such as General Mark Clark, UN Commander in Korea, MajGen James C. Fry, Commanding General, U.S. Army, 2nd Division, LtGen Reuben E. Jenkins, Commanding General, IX Corps, and MajGen George W. Smythe, Commanding General, U.S. Army 3rd Division. He also had chow with troops of the 1st Battalion, 15th Infantry Regiment, which he had commanded at Fort Lewis, Washington in 1940, flew in a reconnaissance plane along the 38th Parallel, met with South Korean President Syngman Rhee--and said no to the use of atomic weapons.

Historian Jean Edward Smith revealed that Ike had a

conversation with General Clark about what it would take to win the war (Smith 559). Clark and General James Van Fleet, commander of the U.S. Eighth Army, had concocted a plan that involved reinforcing the U.S. Eighth Army with eight divisions, launching air and naval operations against the Chinese mainland, and integrating Chinese Nationalist troops from Formosa into the UN operations.

Clark also proposed that the UN give serious consideration to the use of atomic weapons. Some of Clark's ideas were reminiscent of those espoused by MacArthur earlier in the war. Truman had rejected them then. Ike rejected them in 1952. The major difference was that Clark retained his job; MacArthur lost his.

A Side Trip To North Korea?

Some people were sure the president spoke directly to the North Koreans and threatened them with nuclear weapons, although there is no record of such meetings. More than likely, he would not have had time to slip away during his three-day trip without being noticed.

One person who suggests that he spoke with North Korean leaders is Mike Morra, who described Ike's daring trip to North Korea in a letter published on p. 65 of the Sept/Oct 2014 edition of The Graybeards:

"There is a part of history that Korean War veterans ought to know. It has to do with why many Korean War veterans today are still alive or escaped being wounded. During the first few months of 1953, President Dwight Eisenhower flew secretly to North Korea, under the protection of the 82nd Airborne. He faced down the arrogant North Korean generals and warned them, 'This war is over or I'm going to nuke you.' Within a few months all combat ceased.

"General Eisenhower's incredible leadership ended the shooting war. However, if North Korea had reneged, as military assurance the 25th Division, an elite combat unit stationed at Schofield Barracks in Hawaii, was combat ready to be thrown into battle within 24 hours' notice. In gratitude, I recommend that Korean War veterans visit the Eisenhower Library in Abilene, Kansas when they have the chance."

No More From Morra

The fact that Morra did not present any evidence to support his claim is salient. And, the trip to which he alluded would have taken place much later than the one documented in 1952. Predictably, some Korean War veterans challenged Morra's assertion. J.E. Phinazee responded:

"On p. 65 of the September–October 2014 issue [of The Graybeards] there was a letter titled 'A Daring Trip by Ike.' The writer stated that in the spring of 1953, President Eisenhower made a secret trip to North Korea, protected by the 82nd Airborne, where he confronted the North Korean generals at the Pyongyang Peace negotiations, and threatened to "nuke" them, possibly being the reason for the end to the fighting in July 1953.

"Neither I, nor any of my Korean War Veterans Association friends, have ever heard about such a trip, or such personal involvement by President Eisenhower. Since I would not think the letter would have been printed without some verifying information, I would like to know if it was accompanied by reference to any source material that might be used to obtain further information." (Note: Morra writer did not provide any such evidence. And, Ike visited Korea in the winter of 1952, not the spring of 1953.)

You Have To Talk To The Chinese, Not The North Koreans

Robert Hall, a U.S. Marine veteran of the war, expressed similar doubts. He noted that "Mike Morra implies that President-elect Eisenhower, under the protection of the 82nd Airborne, made a secret trip to Korea and faced down the North Korean generals, threatening them with the use of the atom bomb. I can't find a scintilla of evidence that such a meeting took place," he averred. "Both the Truman and Ike's administrations dropped hints that it was under consideration, however."

Hall explained that "Ike would have had to deal with the Chinese anyhow because they were our principal adversaries. His mission was even more secret than the usual presidential trip, but he had announced that he planned to go."

Hall conceded that Eisenhower's election had a positive effect on the continuance of peace talks. "Ike's election no doubt encouraged the communists to return to the bargaining table, but other developments were probably just as influential." He mentioned "Annie" as one of those developments.

"The death of Stalin in March 1953 is considered important by several historians. Others think that the leaking of information about the development of a 280mm cannon capable of firing a tactical atomic projectile could have a key role in the psychological game." At least one historian, T. R. Fehrenbach, a Korean War veteran himself, agreed with the importance of the 280mm cannon.

Most of all Hall decried Morra's charge about the 25th Division and its role in Ike's visit. As he said, "But the suggestion that the threat of deploying the 25th Division, an elite combat unit based in Hawaii, would intimidate our communist adversaries, who ate ROK (Republic of Korea) divisions for breakfast, is both naïve and ludicrous."

And, Hall concluded, "According to Wikipedia, [the 25th Division] was already on the Main Line of Resistance at the time of Ike's visit." Apparently, the truth about the impact of Ike's visit to Korea could not be determined precisely—or even imprecisely.

Actually, the 25th Division was in Korea almost from the start of the war. According to the division's history, it was on occupation duty on the Japanese island of Honshu when the North Koreans invaded South Korea. Commanders were advised during the first week in July 1950 to prepare to deploy to Korea. An advance party from the division landed at Pusan on 8 July 1950. The division headquarters arrived at Pusan four days later. By 18 July the entire division was in Korea. There it remained until the cease fire took effect on 27 July 1953.

Retired U.S. Army Lieutenant Colonel Donald C. Cook also presented ample evidence refuting Morra's claim.

"Ike" Never Got To North Korea

"I recognize that after 65 years memories fade and mistakes are made. I read the issues from cover to cover and just ignore the minor discrepancies which are inevitable, but two letters caught my attention and I felt I had to respond to them.

"The first letter is Bob Hall's letter in the March-April issue. I totally support his response to Mike Morra's letter about "Ike's" trip to Korea. When I read [Morra's] letter, I found the story of Ike's trip to North Korea totally unbelievable. Would Ike go into an enemy country where a state of war still existed and where it would be impossible to protect him? Forget the threat of a division. They wouldn't be there with him, so what could they do? This called for some research, which I did.

"Ike did make a secret trip to Korea. It was kept secret to prevent the possibility of an assassination. He arrived on Dec. 2, 1952 and brought journalists and photographers with him and proceeded to do what you would expect of him, i.e., visiting the front and the troops and commanders to get an assessment of the situation. It's all fully documented.

"The North was obviously not intimidated by his presence, because within an hour of his departure on the 5th from Seoul Airport, the airport was raided by eleven enemy fighters.

"So where was the 82nd Airborne during this? Since WW2 the 82nd had been held in strategic reserve in the U.S. by Truman-- and later Ike--in the event of an attack anywhere by the Soviet Union.

"And where was the 25th? Right where they had been all along--in Korea. They spent 37 months in Korea during the war and did not get re-deployed to Schofield Barracks until September-October 1954. 'Nuff' said!"

Was Morra Wrong?

Morra's critics offered valid criticisms of his assertion that Ike spoke directly with the North Koreans. And, the lack of direct evidence they offer for such a conference is telling. But, based on the total secrecy surrounding Ike's trip from the beginning, it is conceivable that he might have deviated from his agenda in Korea.

The entire trip was cloaked in secrecy from the beginning on 29 November 1952. Here is a truncated point-by-point description of the morning on which Ike left (Goulden 624):

- The personnel going on the trip stood in the pre-dawn dark at scattered locations throughout Manhattan
- Limousines picked them up at their locations
- Journalists and photographers all sworn to secrecy about the trip assembled at Penn Station and exited via a ramp used by mail trucks
- A Secret Service agent drew a police officer near Ike's residence into a conference on an imaginary subject to distract him as the president elect left his residence
- During the day a parade of prominent people entered and left Ike's home to create the impression that he was there conducting business as usual

The secrecy plan continued in Korea, since it was a war zone and Ike could become a target of the Communists. Overall, the trip was so well planned and masked that very few people knew that he was in Korea until after he returned to the United States on 5 December 1952. As George M. Humphrey, who Ike appointed

Secretary of the Treasury in 1953, said afterwards, the entire operation was the "most hush-hush, cloak and dagger...you ever heard in your life."

Given all the secrecy surrounding his visit, and the misinformation and ruses distributed and employed by the planners and protection personnel, it is not out of the question that Ike could have slipped away in Korea and met with high-level communist leaders. However, the idea that he did is pure speculation and cannon fodder (conventional, not atomic, cannon, of course) for conspiracy theorists. Then again, Morra may have been right.

Ike Did Not End The War

The back and forth was typical of the mystery surrounding exactly what President Eisenhower did on his trip to Korea and how it affected the outcome of the war which is technically still in progress. (Significantly, the South Koreans accepted the cease fire under duress, but they never signed any documents to acknowledge that fact.)

The North Koreans have violated that cease fire several times since 1953, and armed guards from South Korea and North Korea still glare at one another at the DMZ (demilitarized zone) in Panmunjom, ready to shoot one another at the slightest provocation. The North Koreans have accused the UN of violating the agreement as well. Charges and counter-charges aside, technically, the war goes on, despite Eisenhower's efforts to end it.

Eisenhower's resolve to end the war using any means available was never in doubt. He made a promise to the American people and he intended to keep it, knowing full well that they were neither 100 percent behind the war—or anywhere near 100 percent—nor willing to use nuclear weapons to attain an end to the fighting. Ironically, it is the North Koreans now who constantly threaten to use nuclear weapons against their sworn enemies. How times have changed!

Chapter 14
Do We Have To Drop Another A-Bomb?

"I have to bring to your notice a terrifying reality: with the development of nuclear weapons Man has acquired, for the first time in history, the technical means to destroy the whole of civilization in a single act." Joseph Rotblat

There was some question during the Korean War as to whether the American public would accept the use of nuclear weapons again. Some of the other UN countries involved in the war would not. President Truman had mixed feelings about deploying them, since he did not want to be remembered by the world for whatever destruction resulted from his use of nuclear weapons for the second time. Nevertheless, in December 1950 he recommended their deployment.

The British would not go along with the suggestion and they convinced their UN allies to argue against their application as well (Korea Reborn 65). Prime Minister Atlee of the United Kingdom made a special trip to the U.S. to clarify the president's stance on the use of nuclear weapons in Korea.

At a 30 November 1950 press conference, President Truman said that the U.S. would use any means at its disposal to end the war. That same day, Air Force General George Stratemeyer advised his counterpart General Hoyt S. Vandenberg that the Strategic Air Command should be ready "to dispatch without delay medium bomb groups to the Far East . . . this augmentation should include atomic capabilities." Stratemeyer's order no doubt created a moral dilemma for Vandenberg, who was not a strong supporter of the war at the onset. He changed his mind and tactics as the stalemate persisted and casualties increased.

Vandenberg's vacillation about the use of nuclear weapons was not unusual among military leaders in Korea. Many of them had served in WWII, and they were used to carte blanche when it came to applying tactics, strategies, weapons, troops, etc. Their ability to make such decisions in the Korean War was hampered considerably by political interference, which was new to them—and to the military in general.

It was tough for the old "warhorses" to adjust to the new military-political relationship, the fact that the U.S. government seemed more intent on winning a political, rather than a military, victory in Korea, and their limited ability to make decisions on their own. That was a dilemma for Vandenberg and his counterparts in the other services.

The General Morphs Into A Hawk

By April 1952 Vandenberg had become a "hawk." He advocated bombing previously off-limit targets such as hydroelectric plants, oil refineries, and dams. A year later he sided with the JCS when it recommended the extensive use of strategic and tactical nuclear weapons to end the war. He did not get to see his recommendation applied.

Vandenberg retired on 30 June 1953, four weeks before the ceasefire was signed. But his vacillation concerning the use of nuclear weapons was not unusual among political and military leaders of the time. The results of the 30 November press conference were proof of that.

It is worth noting that Truman's daughter Margaret believed that the reporters at the meeting, including Tony Leviero of the New York Times, Jack Dougherty of the New York Daily News, Paul R. Leach of the Chicago Daily News, Merriman Smith of the

United Press, and others boxed the president into several corners regarding the use of the atomic bomb. Then the wire services deliberately printed misleading statements about his intentions.

The United Press ignited the controversy with these words: "President Truman said today the United States has under consideration use of the atomic bomb in connection with the war in Korea."

The Associated Press (AP) phrased its analysis of the press conference a little differently: "President Truman said today active consideration is being given to use of the atomic bomb against the Chinese Communists if that step is necessary." Later, the AP clarified its first distortion with another distortion, claiming the president said, "The decision of whether to drop atomic bombs was one for the commander in the field."

Finally, the AP reported, "President Truman said today use of the atomic bomb in Korea has always been under consideration--and whether it is used is up to American military leaders in the field." The subtle differences in each ensuing printed interpretation of each statement only succeeded in confusing the American public and U.S. allies.

What Actually Happened At The Conference?

At the gathering a reporter asked Truman if his statement about "any means" included nuclear weapons. The president replied that all weapons were under consideration. But, he added, "There has always been active consideration of its use. I don't want to see it used. It is a terrible weapon" (Haruki 150). His comment created a stir among some of the UN allies.

The president felt compelled to clarify his statement. His press secretary, Charlie Ross, explained that according to the U.S. Constitution only the president could authorize the use of nuclear weapons. (Exactly where in the Constitution that is written is unclear.) Once he did it was up to the commanders in the field to decide when and where they would be used (Catchpole 97). That begged the question. It still did not clarify whether the president would okay employing nuclear weapons.

Reporters assumed the clarification meant he would and they let the world know. That did not set well with many leaders around the world—especially the British, the French, and the Canadians.* Once again the press did not help. One Italian newspaper published a story that bombers on Japanese airfields were loaded with atomic bombs and ready to take off at a moment's notice. The Times of India ran a headline over an editorial saying in large letters "NO, NO, NO" (Truman 497-98).

* Ironically, the Canadians were prime contributors to the development of the atomic bomb. Canada was one of only three sources of uranium at the time the U.S.'s bomb was being developed. The other two were the Belgian Congo and Czechoslovakia (the Jachymov mine in Bohemia), from which Russia acquired its uranium. It wasn't until after 1945 that uranium was found in the USSR and other Central Europe locations

The firestorm created by the reporters and the clarifications to the clarifications created a need for additional clarification explaining the president's real intention regarding the use of atomic weapons in Korea. Atlee appointed himself as the person to discuss the situation directly with President Truman, despite the president's assurance to his allies that there was no need for any conferences.

With apologies to Dr. Seuss and Horton the elephant, Truman believed he had said what he said and meant what he said. But, reporters were not exactly sure what that was. (One of the classic quotes from Dr. Seuss's book Horton Hears A Who is "I meant what I said and I said what I meant. An elephant's faithful one hundred percent.")

There's Never A Dulles Moment In The White House

Atlee flew to the "colonies" in early December 1950, just

two months after the Chinese had entered Korea, and began pushing UN forces back to the 38th Parallel. It was not a good time militarily for the UN. Nevertheless, Atlee and his supporters were not eager to have the U.S. introduce nuclear weapons against the Communists—especially in China—and possibly start WWIII. They had reason for concern.

There was some evidence that the U.S. military was planning to rely more on its nuclear weapons arsenal after WWII and less on conventional warfare. That was borne out by a famous statement by Secretary of Defense Louis Johnson after he cut $12,333,294,000 from the armed forces' budget for the fiscal year 1951. Johnson said, "If the Russians hit America at 4 a.m., America's atomic bombers would strike back by 5 a.m." (Blair 27).

President Eisenhower's Secretary of State, John Foster Dulles, a former U.S. Army veteran and U.S. Senator, was of a like mind. He espoused a policy of "brinksmanship," which he credited for ending the war in Korea. (A generally accepted definition of the term is "the ability to get to the verge without getting into a war.") Dulles looked at the ability to play that game as a "necessary art."

A la MacArthur, Johnson was relieved of his duties by Truman. Dulles, who believed that massive retaliation was an important part of "brinkmanship," stayed on. As he explained in a 12 January 1954 speech, the president and National Security Council had decided "to depend primarily upon a great capacity to retaliate instantly, by means and at places of our own choosing" (Dulles, Arlington).

He explained later that his skill in playing "brinksmanship" had ended the Korean War. That is evident based on a paragraph in the "Short History of the Department of State:"

"In a 1956 Life magazine interview, Dulles described how he had passed the word to the Chinese and the North Koreans that unless the communist powers signed the Korean armistice, the United States would unleash its atomic arsenal. Dulles claimed that by moving to the brink of atomic war, he ended the Korean War and avoided a larger conflict. From that point on, Dulles was associated with the concepts of "massive retaliation" and "brinksmanship," a supposedly reckless combination of atomic saber rattling and eyeball-to-eyeball standoffs."

In reality, then, the so-called atomic threat to China was less definitive than Dulles had claimed, and the Eisenhower Administration policy of "massive retaliation" was far more cautiously based on mutual atomic deterrence. Dulles claimed that the implied threat to drop atomic bombs on Manchuria had prompted the enemy to say enough was enough. He wasn't the first person to say that. He believed in striking first with nuclear weapons. The question was what the Russians would do in response.

U.S. military and political leaders knew that Russia might retaliate in kind, because it had conducted its first test of an atomic bomb on 29 August 1949. They weren't sure how far along the Russian program was, but they didn't want to take any chances. The U.S. was developing and testing more powerful weapons than they had in their current arsenal, including a hydrogen bomb. That explains in part why the allies were nervous about the use of nuclear weapons in Korea.

Stalin was proud of Russia's nuclear weapon development program, and he took a page out of the U.S. military's book by using it as a psychological weapon. He told North Korean leader Kim Il Sung in 1950 that news of the Russian program was unsettling to the American public, which might dissuade their government from intervening in Korea (Haruki 58). Stalin was wrong. (Actually the Russians did not produce any serviceable nuclear weapons until 1954, too late to be used in Korea.)

Atlee and Truman held a series of face-to-face meetings over a period of several days in early December on a variety of war-related topics. There was fuel added to the fire just before their talks began on 4 December when South Korean Defense Minister Shin Sung Mo said that his countrymen would rather die in an atomic bomb blast than become Chinese slaves.

Therefore, Mo said, the use of nuclear weapons would be a good thing if it would bring an end to the war—and the quicker the better. That was not what Atlee and Truman wanted to hear.

Splitting Atoms, Splitting Hairs

Just before Atlee and Truman met, U.S. Army Chief of Staff General Joseph L. Collins was in Korea on a fact finding mission. Collins was not an advocate of the use of nuclear weapons, but military and political leaders valued his opinion nonetheless. Advisors decided to wait for Collins' report on the situation there before they made any decisions about what Truman could tell Atlee in their upcoming talks, which started in a rigid fashion and ended with no concrete results.

According to the 5 December 1950 minutes of the first meetings, "The secretary said he had suggested to [British] Ambassador [Oliver S.] Franks that he attempt to work out with Mr. Atlee a means of getting a more relaxed attitude at future meetings with the President...The meeting which was held at 4 p.m. yesterday (December 4) was rather rigid and too many people were in attendance." The environment was understandable. Atlee "had taken the position that at this time we had no choice except to negotiate with the Chinese." The Americans did not agree.

The talks dragged on, but the two leaders did not get around to settling the nuclear weapons issue immediately. Ironically, they started discussing the use of nuclear weapons on 7 December 1950, the ninth anniversary of the Japanese attack on Pearl Harbor. The results were favorable for Atlee—with a caveat.

The president promised Atlee that he would not authorize the use of nuclear weapons without getting approval first from the British prime minister. Truman was playing word games. Later the promise was interpreted to mean that he was referring to the prime minister at the time, i.e., Atlee, and not his successors.

In a joint statement issued by the representatives of the two countries after they had completed six meetings, there was one reference to atomic bombs:

"The President stated that it was his hope that world conditions would never call for the use of the atomic bomb. The President told the Prime Minister that it was also his desire to keep the Prime Minister at all times informed of developments which might bring about a change in the situation."

Fortunately, the semantical hair splitting never came to a head. That settled the conversation between Truman and Atlee. But the discussion of the use of nuclear weapons arose again between Truman and MacArthur and when Ike succeeded Truman.

Chapter 15
Harry And "Mac" Differ On The Attack

"We had news this morning of another successful atomic bomb being dropped on Nagasaki. These two heavy blows have fallen in quick succession upon the Japanese and there will be quite a little space before we intend to drop another." Henry L. Stimson

In April 1951 President Truman fired General MacArthur, much to the consternation of American citizens, politicians, and military leaders. Conventional history suggests that Truman was dissatisfied with the general's poor military leadership, which is why he replaced him with General Matthew Ridgway. Often lost in the historical accounts are the two leaders' divergent views on the use of nuclear weapons.

MacArthur believed that the use of nuclear weapons would end the war because the radioactive fallout zones would interrupt the Chinese supply zones. He didn't want to inundate North Korea with nuclear bombs. All he wanted was a couple dozen to get their attention. The problem was that the U.S. didn't have enough nuclear bombs in its arsenal to use in Korea without upsetting the balance needed to keep some on hand for use if other enemies (Russia) got rambunctious. That was due in part to President Truman's lack of oversight of the nation's post-WWII nuclear weapons programs, which carried over into the Korean War.

When MacArthur Got There, The Cupboard Was Bare

Following WWII President Truman had a lot on his mind. His most pressing problems were to help rebuild the devastated countries the U.S. and its allies had defeated and to engage in the time-honored practice of downsizing the military following a war. Included in the latter activity was cutting back on the stock of nuclear bombs in the U.S.'s arsenal.

There were several reasons the nation's nuclear stockpile dwindled. One was the public's distaste for nuclear warfare. Even though the use of the atomic bombs over Hiroshima and Nagasaki hastened Japan's willingness to surrender, there were segments of the military and the public that were less than enthusiastic about developing a nuclear strategy or building new nuclear weapons to put in service. Another was a lack of plutonium production following the war. And, the farthest thing from the U.S. government's mind was getting involved in another war anytime soon. So, addressing the stockpile of nuclear bombs took a back seat to other problems.

By April 1948 the U.S. had approximately twelve nuclear bombs to its name, all of which were unassembled. The assembly process required 24 men working for two days to ready the bombs for combat. Strangely enough, the Air Force had 32 B-29 bombers available to deliver them if it became necessary (Blair 9-10)-- almost three times more than it needed. Where and when that would be was a bit fuzzy.

The U.S. had only one major enemy in mind at the time. That was Russia. The two countries were already engaged in a Cold War, but nobody on either side anticipated a nuclear war anytime soon. The Russians were in the process of developing their nuclear arsenal. The U.S. had developed and used nuclear bombs successfully, but did not show much interest in expanding their use. Something had to give.

Either Truman Was Apathetic Or He Just Didn't Care

Truman paid scant attention to the U.S. nuclear weapons strategy and development during the first two years of his presidency. (He had succeeded President Franklin Delano Roosevelt, who died in office on 1 April 1945.) He tried to foist management of all nuclear energy programs onto the UN and oversaw the creation of an Atomic Energy Commission, which transferred the control of atomic energy from military to civilian authority. Beyond that, he didn't seem to care about how many nuclear bombs the U.S. had in its cupboard, what new developments in nuclear energy were taking place around the world, or what its future held. Yet, he professed to be shocked in 1947 when he learned just how few nuclear bombs the U.S. had.

Truman may not have been concerned about the dearth of bombs, but the Pentagon was. In 1947 and 1948 Pentagon officials pleaded with their civilian counterparts to raise the number of available nuclear weapons to at least 400. Their pleas fell on deaf ears. They did, however, get the number of available nuclear bombs increased to fifty by mid-1948—all of them "old technology" weapons. Then, things began to change.

By December 1948 scientists at the Las Alamos Scientific Laboratory announced that they had developed a new and improved type of nuclear weapon that was more efficient and required less fissionable material. There didn't seem to be any hurry to develop the new weapons. Instead, the U.S. added another fifty bombs of the Nagasaki-era type.

In their infinite wisdom, planners made available 100 bombers to drop their 100 bombs. It didn't seem to matter to a lot of people other than U.S. Navy personnel. There was no war imminent and no one to drop the bombs on anyway.

The Navy Revolts

It appeared to some administrators in the U.S. Navy as 1950 approached that the U.S. was developing a policy of reliance on nuclear weapons as the keystone of its mid-20th century military strategy and that the Air Force was favored as the delivery service.

They believed that the lack of advances in nuclear development did not support that strategy. Some went so far as to proclaim nuclear warfare immoral.

The administrators alleged among other things that reliance on nuclear warfare would neither defeat nor deter an enemy (read Russia), nuclear weapons were not reliable, the plane designated to carry them, the B-36, was not capable of reaching some targets in Russia, the drops would not be accurate, the weapons were not as powerful as they were advertised to be...the list went on. The JCS refuted all these allegations, many of which were aimed at the wrong country as a potential target.

Lo and behold, when the next war started it was not in Russia; it was in Korea. And it was not against Russia; it was against North Korea and China (although Russia played a significant behind-the-scenes support role of the other two communist countries). Furthermore, the arguments over whether to use nuclear weapons was still unresolved.

The U.S. continued to develop new ones, e.g., "Annie," but there was no consensus on how and when to use them and the American public was ambivalent about the issue. The argument played a big role in Truman's decision to axe MacArthur, and continued when Eisenhower replaced Truman.

MacArthur's Side

After hearing what Truman said about field commanders' authority to use nuclear bombs, MacArthur tried to take advantage of the comment. On 9 December 1950 he asked JCS's permission to employ nuclear bombs. Fifteen days later he asked for 26 bombs and provided a list of targets for them. He requested four more to drop on the "invasion forces," i.e., the Chinese, and another four for "critical concentrations of enemy air power."

It wasn't until 1964 that MacArthur's plans for the bombs were revealed. They appeared in the aforementioned 9 April 1964

New York Times article published a week after his death.

"The enemy's airpower would first have been taken out. I would have dropped between 30 and 50 atomic bombs on his airbases and other depots strung across the neck of Manchuria from just across the Yalu River from Antung to Hunchun. Between 30 and 50 atomic bombs would have more than done the job. Dropped under cover of darkness they would have destroyed the enemy's air force on the ground, wiped out his maintenance and his airmen."

Two of his statements were at odds with one another. If enemy airpower had been taken out, it would not have mattered whether the bombs were dropped at night or during the day, except to the U.S. air crews who would in all probability have to stave off more enemy planes in the daytime. The primary advantage to dropping them "under the cover of darkness" would have been psychological. That certainly wouldn't have mattered if the bombers delivered between 30 and 50 atomic bombs. There wouldn't have been enough enemies alive to worry about the psychology of the raids. Ah, just more fodder for the speculation barrel.

Speculation aside, MacArthur's December 9th request wasn't the first time he had asked for atomic bombs. His first request came only two weeks after the U.S. entered the war.

Please, Sir: More Bombs

On 9 July 1950 the JCS considered whether MacArthur should have access to atomic bombs. General Charles Lawrence Bolte, who commanded the 7th Army in West Germany and was Army Commander-in-Chief in Europe in 1952-3, was assigned to discuss their use with MacArthur. Bolte felt that the request was feasible, and that the U.S. could provide between 10-20 bombs.

After careful consideration, the JCS decided not to send MacArthur the bombs. They just didn't think that the targets

MacArthur was proposing required the use of such large weapons. MacArthur shelved his request until the following December. That no doubt pleased Truman.

Truman simply did not trust MacArthur when it came to the use of atomic bombs. Oddly enough, it wasn't necessarily MacArthur he had to worry about. There were other military leaders and politicians who supported their use or offered bizarre ideas.

Air Force General Curtis LeMay was willing to direct a nuclear bomb attack campaign attack against Communist China. He was not concerned with the morality or immorality of using nuclear weapons. As he said, "Every soldier thinks something of the moral aspects of what he is doing. But all war is immoral and if you let that bother you, you're not a good soldier."

Congressman Albert Gore Sr. suggested that the UN create a radiation belt dividing the Korean peninsula permanently into two. Perhaps Truman would have been better off worrying about plans like Lemay's and Gore's instead of MacArthur's.

LeMay and Gore aside, MacArthur never got the chance to utilize nuclear weapons in Korea, nor did anyone else. Truman relieved him of his duties in April 1951, partly because of his unwillingness to trust the general with nuclear weapons. That was shortly after 10 March 1951, when MacArthur requested a "D-Day atomic capability." The timing could not have been coincidental.

Chapter 16
The Chinese Are Coming En Masse

"Today I can declare my hope and declare it from the bottom of my heart that we will eventually see the time when that number of nuclear weapons is down to zero and the world is a much better place." Colin Powell

By late 1950 the Chinese were amassing troops near the Korean border to begin a new offensive and the Russians were staging 200 bombers at Manchurian air bases that gave them the capability of hitting targets in Korea and Japan. Oddly enough, MacArthur and his intelligence staff refused to acknowledge or shrugged off the reports that as many as 300,000 Chinese troops had crossed the frozen-over Yalu River at night and moved into North Korea with the intention of annihilating UN forces in the area.

Despite those circumstances, MacArthur still did not get the nuclear bombs he wanted—but his request apparently hastened his removal. That was not publicized as part of the equation at the time. Contemporary historians have advanced that theory.

Bruce Cumings epitomized that school of thought in a History Channel article:

"The US came closest to using atomic weapons in April 1951, when Truman removed MacArthur. Although much related to this episode is still classified, it is now clear that Truman did not remove MacArthur simply because of his repeated insubordination, but because he wanted a reliable commander on the scene should Washington decide to use nuclear weapons; Truman traded MacArthur for his atomic policies."

A Bit Short Of A Whole Kit And Kaboodle At Kadena

Even while Truman was sacking MacArthur in part because of his willingness to use nuclear weapons, the U.S. was stockpiling them at Kadena Air Force Base on Okinawa. The Chinese and Russian deployments of troops and bombers near the Manchurian border concerned Truman and his military leaders, even if MacArthur and his staff disregarded them. They warned the Chinese not to cross the border, and leaked the news that they would utilize nuclear weapons if they did.

Just for good measure, they assigned crews at Kadena to start assembling nuclear bombs. They did not include the nuclear cores in the process, though. That vital step was deferred until someone decided to actually use the bombs.

Truman's Thoughts

President Truman was all over the map in his thoughts about the use of nuclear weapons in Korea. He said at some points that they might be employed, they might be helpful…but he never got around to ordering their use. As his discussions with Atlee demonstrated, he did not make his allies happy when it came to making decisions about nuclear weapons.

The president of South Korea, Syngman Rhee, was disappointed in Truman's cautious approach. Rhee admitted that nuclear weapons were horrible, but he saw their use as a beneficial tool in repelling his adversaries, especially after the Chinese entered the war and threatened to turn its tide in favor of the Communists. One Korean native who came to the U.S. after the war agreed with Rhee, but he stirred up a bit of controversy in the process.

On The Lee Side, But Not Safely

Hubert Hojae Lee, a renowned economist who was an elementary school student when the war began, wrote in the

Jan/Feb 2011 edition of The Graybeards (p. 60) that President Truman erred by not acceding to MacArthur's plans to use the atomic bomb in Korea. (Read his comments in his book, My Journey To America, p. 136.) His assertion did not go unchallenged.

Lee wrote:

"Many historians and pundits are still debating the wisdom of MacArthur's intention to use an atomic bomb to end the Korean War. General Alexander Haig, Jr., confirmed that intention when I had an interview with him several years ago. Most Korean people believe that President Harry Truman's myopic view permanently divided Korea into the two Koreas we have today. If General MacArthur had been allowed to carry out his military strategy to smash the communists by using the bomb, there would be only one Korea today.

"Imagine the waste of limited economic and human resources on both America and South Korea for the past 60 years, especially the sufferings and sadness of many divided Korean families in that time. There is strong justification on the side of Korean people for General MacArthur's historical vision in asserting that 'There is No Substitute for Victory in the Korean War,' though there would have been a serious fallout following the use of an atomic bomb.

"I believe that the benefit of a united Korea would far outweigh the cost of the fallout.

The historical mistake made by Harry Truman will continue to plague America in the future in terms of Far East Asian security and a balance of power, as China emerges ever stronger—enough to be threatening the deeper involvement by the United States of America with an economic stability and military security in the Korean Peninsula."

Some veterans did not agree with Lee. U.S. Marine Corps veteran Robert Hall responded:

"Dr. Hubert Lee reopened a dormant can of worms when he contended that the Koreans blame Pres. Truman for preventing the unification of Korea. Absent an actual poll, I believe this is mostly a reflection of Dr. Lee's bias.

"'If only Gen. MacArthur had been allowed to carry out his military strategy to smash the communists by using the bomb (emphasis mine),' he says, 'there would be only one Korea today.' ("The bomb" can only mean the atom bomb.)

"As I see it, if Truman had not acted without hesitation, as he did in responding to the North Korean invasion of June 25, 1950, Koreans would be living in a unified country, but they would be virtual slaves, like their countrymen in North Korea today, under a criminal communist regime.

"Let's examine MacArthur's grand strategy. He sent UN forces on their way to the Yalu despite repeated warnings from Red China that it would trigger its intervention. He advocated bombing China's industrial centers, blockading China's coast, assisting Chiang Kai-Shek to invade the Chinese mainland, and even bolstering UN forces with some of Chiang's men.

Additionally, MacArthur sought carte blanche use of the atom bomb in North Korea to destroy its industrial base and even to lay a field of radioactive wastes across the enemy's supply lines.

"General Ridgway, who whipped a demoralized Eighth Army into shape after the Chinese appeared to be invincible, and who was an admirer of MacArthur, says in his book The Korean War that had not Truman fired MacArthur he would have been "derelict in his own duty."

"Truman, Secretary of Defense Marshall, the JCS, and our allies were unanimous in their belief that MacArthur's agenda was not only impracticable but carried the risk of bringing the Soviet Union into the war.

"MacArthur cannot be denied credit for at least two spectacular successes: the [September 1950] Inchon invasion, which relieved pressure on the Pusan perimeter, and the manner in which he put a devastated Japan on a democratic course to become a modern industrial power.

"But MacArthur's downfall resulted from his arrogance and pomposity and his conviction that he was infallible. He may have been a demigod to the Japanese, but to the administration and those responsible for our armed forces he suffered from delusions of grandeur that led to his frequent acts of insubordination.

"Dr. Lee undoubtedly reflects the views of Syngman Rhee (long since deceased). But Truman hardly deserves vilification by the South Koreans. On the contrary, there should be statues to our former president throughout the republic

Another veteran, Robert W. Robinson, took them both to task:

"The discussion in the March/April issue of The Graybeards ("Were There Worms In "The Bomb?" p. 67) omitted a couple vital points that led to the war and its conduct.

1. Russia was in it from the time Stalin gave Kim the green light to attack the south.

2. MacArthur's battle plans were funneled to the North Koreans, China, and the USSR through the United Nations. One can never win with that handicap.

"So much for 'dereliction of duty,' egoism, etc. Many of us appreciate a commander on the same side of the fight."

And so the argument raged—six decades after the fighting in Korea ended. The fact remains that Truman fired MacArthur, the bombs were never dropped, and the two Koreas remain divided. Hindsight will not resolve the issue of what impact the use or non-use of nuclear weapons in Korea would have had—and it never will.

The Handoff To Ike

Rhee's disappointment aside, Truman's unwillingness to

authorize the use of nuclear weapons carried over past MacArthur's firing, with one exception.

By April of 1951 the JCS was concerned with a Chinese troop build-up and the possibility that it would include the use of bombers to attack U.S. troops and bases. The JCS requested that nuclear weapons be transferred from the AEC's control to the military. Truman approved the request at just about the same time he relieved MacArthur. Somehow his order to transfer the weapons got lost in the shuffle.

As the war dragged on so did the number of threats emanating from Washington DC to use nuclear weapons. They were serious enough to make Chinese, North Korean, and Russian leaders sit up and take notice. But the threats were never translated into action, and Truman never had to make a final decision about nuclear weapon use. By the time the 1952 presidential campaign began, the decision became Eisenhower's problem.

Chapter 17
We Like "Ike," But We Are Tired Of The War

"I do not believe that civilization will be wiped out in a war fought with the atomic bomb. Perhaps two-thirds of the people of the earth will be killed." Albert Einstein

After President Eisenhower was elected in 1952, seven+ years had passed since the atomic bombs were first dropped in warfare. Nevertheless, the effects of and arguments about the use of such weapons were still fresh in Americans' minds. And, to make matters worse for Ike, the Korean War was not particularly popular in the United States, which had just recently finished its participation in WWII. Americans were tired of war.

Popular Opinion Shows That The War Is Not Popular

According to a 2 June 1953 White House memo summarizing polls on Korea, by May 1953, almost three years into the war, only seventeen percent of the American population believed that the U.S. would be able to reach some kind of peace agreement with the Communists in the next month. That was down from 34 percent only a month earlier, when 69 percent of the population would have welcomed a truce—any truce.

Even if there were some kind of a truce signed it would not be a good thing according to the polls. Only 45 percent of the respondents acknowledged that signing a truce would mean that the UN had completed its mission successfully. Perhaps the most telling outcome of the polling was the percentage of people who did not believe that the war was worth fighting in the first place.

The question was: "As things stand now, do you feel that the war in Korea has been worth fighting, or not?" The percentages remained fairly consistent.

	Oct. 1952	Nov. 1952	Jan. 1953	Apr. 1953
Worth fighting	32%	34%	39%	36%
Not worth fighting	56%	52%	58%	55%
No opinion	12%	8%	9%	9%
	100%	100%	100%	100%

Nevertheless, 62 percent of Americans were willing to take strong steps to end the war, even though "our allies in the United Nations refuse to go along with us." That matter was addressed between Eisenhower and Dulles at a National Security Council meeting on 11 February 1953, as reported by Bernard Gwertzman in an 8 June 1984 article in the New York Times. Gwertzman wrote that the president and Dulles agreed that the U.S. could not pursue its Korean strategy indefinitely.

The writer added that at a follow-up meeting on 27 March 1953, Eisenhower and Dulles agreed "that somehow or other the taboo which surrounds the use of atomic weapons would have to be destroyed." Gwertzman also noted that "While Secretary Dulles admitted that in the present state of world opinion, we could not use an A-bomb, we should make every effort now to dissipate this feeling."

Another 31 percent of the respondents did not favor any recommended stronger steps, which included bombing across the Yalu River, employing Chinese Nationalist troops in Korea, and giving their leader Chiang Kai-shek all the support he needed to invade the Chinese mainland. (The Chinese Nationalists based in Formosa, led by Chiang Kai-shek and backed by the U.S., had been at war with the mainland Chinese Communists since 1927.)

Forty-seven percent of the folks polled did not see too great a risk if the U.S. followed those stronger strategies; 41 percent did. In any event, as of June 1953, the poll showed that seventy percent of the U.S. population approved of the way President Eisenhower was handling foreign affairs, particularly his management of the "Korean problem." That support still did not give the president carte blanche to use nuclear weapons, even though he favored their use, in theory at least.

Let's Not Drop The Idea Of Dropping The Bomb

President Eisenhower was not alone in his apparent willingness to employ nuclear weapons, ranging from those dropped from planes to artillery shells. In fact, American military and political leaders had discussed the use of nuclear weapons at the beginning of the war, when their only enemy was North Korea—before China's entrance into the war Their ambivalence about the use of nuclear weapons in Korea reflected U.S. public opinion—even though the question of their use was not included in polls. There was simply no widespread support among the leaders for their use.

Changing leadership, military and political, muddled the waters regarding decisions about the use of nuclear weapons. There did not seem to be any consistency in the decision making process. General MacArthur was a proponent of employing nuclear weapons, especially early in the war, but President Truman had fired him in 1951. So he no longer had any say in the matter. Likewise, President Eisenhower had replaced President Truman after defeating Adlai Stevenson in the 1952 election, so Truman's opinion was moot.

Eisenhower had suggested to then U.S. Army Chief of Staff J. Lawton Collins and General Matthew Ridgway that two atomic bombs should be used in Korea. Collins in particular was against the idea. He explained, "Personally, I am very skeptical about the value of using atomic weapons tactically in Korea. The communists are dug into positions in depth over a front of 150 miles."

And, he added, nuclear tests "proved that men can be very close to the explosion and not be hurt if they are well dug in." (It wasn't until the spring of 1953 that Collins voiced his support for the use of nuclear weapons to bring the war to an end.)

President Eisenhower demurred. He thought "it might be cheaper, dollar wise, to use atomic weapons in Korea than to continue to use conventional weapons against the dugouts which honeycombed the hills along which the enemy forces were presently deployed." He even suggested a target area.

At a National Security Council (NSC) meeting in February 1953, Eisenhower proposed the Kaesong area of North Korea as an appropriate demonstration ground for a tactical nuclear bomb. As he explained, it "provided a good target for this type of weapon."

Recently declassified documents reveal that such recommendations were not particularly rare as the Korean War dragged on. On 19 May 1953, just three months after Eisenhower proposed an actual target, the JSC recommended direct air and naval operations against China, including the use of nuclear weapons. The NSC endorsed the JCS recommendation the following day. None of these decisions would have pleased Ridgway.

Ridgway, who was no fan of the use of nuclear weapons, concurred with Collins. In fact, Ridgway retired from the Army in 1955, partly because he and Eisenhower did not see eye to eye on the use of nuclear weapons or their value as a psychological weapon.

In General The Generals Disagree

Ridgway, who served under Eisenhower in WWII, succeeded General of the Army Douglas MacArthur as commander of United Nations forces in Korea and of allied occupying forces in Japan in

Atomic Cannons and Nuclear Weapons

April 1951 after President Truman axed MacArthur. Then, in June 1952, he replaced General of the Army Dwight D. Eisenhower as Supreme Commander of Allied Forces in Europe. The following year President Eisenhower appointed Ridgway as Army Chief of Staff. That was not because he was a "yes man." Ridgway was not afraid to stand up to Eisenhower, and the two never agreed on the use of nuclear weapons.

Eisenhower did not pull any punches regarding Ridgway's views. He flat out told Ridgway that they were "parochial" because he would not back the new military strategy of using the *threat* of nuclear weapons dropped from aircraft, if not their actual use, even though the practice would allegedly make infantry troops' role safer.

Unlike Ridgway, Eisenhower had no qualms about using threats to employ nuclear weapons. That is what he did in Korea: he dropped threats, not bombs. Eisenhower talked about using nuclear weapons, but never actually ordered their use. Nevertheless, he visited Korea between his election and his inauguration. He concluded as a result of his three-day visit that, "we could not stand forever on a static front and continue to accept casualties without any visible results. Small attacks on small hills would not end this war." Perhaps "Atomic Annie" would.

No one ever found out. The fighting ended before "Annie" arrived in Korea—at least officially—which may or may not have been until 1958. There were eyewitnesses, however, who swore it happened a lot earlier—even during the war. Others aver that "Annie" never got to Korea. There exists no concrete proof one way or another.

Chapter 18
Ike And "Annie" Who?

"The only use for an atomic bomb is to keep somebody else from using one." George Wald

It is interesting to note that historians who wrote the "authoritative" histories of the Korean War were generally ambivalent regarding the significance of the relationship between President Eisenhower and "Atomic Annie" and the outcome of peace talks in Korea—if they mentioned Ike and/or "Annie" at all. Generally, they did not see any relationship, although they mentioned both occasionally.

Many historians, especially those whose works have been published in recent years and have access to more declassified government documents than did their earlier counterparts, created a link between nuclear weapons and the outcome of peace talks, without tying Eisenhower to it. "Annie" was--and still is--rarely mentioned. Most of the references to nuclear weapons were to air-dropped bombs. Eisenhower was a bit player in their histories.

Bruce Cumings and Joseph C. Goulden were two of the rare exceptions to the "air-dropped bombs" approach. Cumings alluded to "Annie" in his aptly named history, The Korean War: "On 26 May 1953 the New York Times featured a story on the first atomic shell shot from a cannon, which exploded at French Flat, Nevada with ten-kiloton force (half the Hiroshima yield)."

Cumings also referenced another blast at the test site a few days later, which might have been a hydrogen bomb. He suggested that stories like that in the New York Times and Eisenhower's veiled threats to use nuclear weapons amounted to "atomic blackmail, a way of getting a message to the enemy that it had better sign the armistice" (Cumings 34). At least "Annie" got some credit from Cumings, albeit indirectly. That was rare among historians.

Goulden cited a "swift rethinking of nuclear policy by the JCS, which had frequently considered the nuclear option the past two years, only to reject the use of atomic bombs as impractical" (Goulden 628). The new philosophy was reflected in a 27 March 1953 passage in a JCS study: "The efficacy of atomic weapons in achieving greater results at less cost of effort in furtherance of U.S. objectives in connection with Korea points to the desirability of re-evaluating the policy which now restricts the use of atomic weapons in the Far East."

Damn The Costs: Full Nuclear Bombs Ahead

The idea of "less cost of effort" was not imbued in all military leaders. The prevailing thought among them was that nuclear weapons alone would not assure victory in Korea—or in any other war. In fact, some said, their use would require more effort on the part of ground troops and lead to higher costs to conduct a war. That idea was highlighted by editor Stephen I. Schwartz in his 1998 U.S. Nuclear Weapons Cost Study Project.

He stressed that "Proponents of the belief that nuclear weapons kept the cold war cold frequently ignore or discount the impact of much larger U.S. expenditures on conventional forces…and the fact that these forces, in contrast to nuclear weapons, were used in actual combat (for example, in Korea and Vietnam)."

Schwartz argued in his study that U.S. leaders never paid much attention to the costs of nuclear warfare from any standpoint, including economic or political. According to his estimates, the U.S. spent $5.8 trillion on nuclear weapons between 1940 and

1998, an excessive amount in his opinion.

The opening to his report stresses the idea of excessive costs: "We will document that the United States spent vast amounts on nuclear weapons without the kind of careful and sustained debate or oversight that are essential both to democratic practice and to sound public policy. In most cases, even rudimentary standards of government policy-making and accountability were lacking. Today, with an estimated $35 billion expended annually on nuclear weapons and weapons-related programs (in 1998 dollars), government officials remain by and large unaware of both the overall size of and rationale for such costs. Notwithstanding frequent debates about arms control treaties, the nature of deterrence, the size and mix of nuclear forces, or the threat these forces were intended to counter, the decision- making process did not allow for a rigorous examination of the costs associated with U.S. nuclear policy. This rendered it politically and fiscally unaccountable."

Schwartz observed that the Korean War played a part in increasing costs for nuclear weapons in the early 1950s. He wrote that an "enormous increase in nuclear budgets in 1951 reflects efforts to respond to the first Soviet atomic test in August 1949, the coming to power of the Communist party in China later that year, and the start of the Korean War in June 1950, events which proved particularly unsettling to civilian policy makers and the military."

It is precisely those factors that led military and political leaders to give serious thought to the use of nuclear weapons in the Korean War, and to beefing up the number of active duty personnel.

At the beginning of the Korean War the U.S. had only 1.5 million men and women on active duty. They were equipped for the most part with WWII-era material. The numbers increased throughout the war. By 1956 there were a projected 2,860,000 men and women in uniform. This almost 100% increase took place even though the U.S. had expanded its nuclear weapons arsenal, which supported some experts' arguments that nuclear warfare would require more—not fewer—ground and support troops. Nevertheless, during the war the JCS was still pushing for the use of nuclear weapons.

The JCS Seemed Eager To Drop The "Big One"

In another passage from the 27 March study the JCS suggested that if the U.S. couldn't develop enough conventional weaponry it might as well resort to nuclear weapons to fill the gap:

"In view of the extensive implications of developing an effective conventional capability in the Far East, the timely use of atomic weapons should be considered against military targets affecting operations in Korea, and operationally planned as an adjunct to any possible military course of action involving direct action against Communist China and Manchuria."

Two months later the JCS pushed forth the idea that the UN Command should warn the Communists that if they didn't report to the peace talks table forthwith it would launch a major offensive against China and Manchuria—including nuclear weapons. In its May 1953 report, the JCS recommended that the operations should include the "extensive strategical and tactical use of atomic bombs."

The National Security Council thought that was a good idea. It approved the JCS's recommendations. Once again the U.S., under the guise of the UN Command, was teetering on the brink of employing nuclear weapons and expanding the war outside Korea. Meanwhile, President Eisenhower was still trying to interest the Communists in peace talks. The snag in reaching an agreement was what to do with prisoners of war, not the use of nuclear weapons.

Leaner, Meaner, Less Intrusive Government Reports

The U.S. State Department's brief history of the war made no reference to President Eisenhower's role in ending the fighting. It focused simply on the main sticking point: the prisoner exchange:

"For the next two years, small-scale skirmishes continued to break out, while the various representatives argued over the peace terms. After agreeing on the demarcation line and the settlement of airfields, the main issue blocking progress in the talks was the repatriation of prisoners of war. On July 27, 1953, the DPRK (North Koreans), PRC (Chinese) and UN signed an armistice (the ROK abstained) agreeing to a new border near the 38th parallel as the demarcation line between North and South Korea.

Both sides would maintain and patrol a demilitarized zone (DMZ) surrounding that boundary line. The armistice also established a commission of neutral nations to oversee the voluntary repatriation of POWs. According to the agreement, each side would have to repatriate willing POWs within sixty days and send unwilling repatriates to the commission to oversee their departure to their preferred destinations. Under the supervision of the commission, some 14,227 Chinese and 7,582 North Koreas opted against repatriation; the Chinese were sent to Taiwan rather than the Chinese mainland.

A handful of U.S. and British POWs in North Korea opted against repatriation as well, choosing instead to live in Communist China or North Korea."

The debate over prisoners' fates did not hinge on the use of nuclear weapons of any type. In fact, the two issues are hardly ever mentioned in the same breath by historians.

Chapter 19
Fehrenbach's History Stands Out

"North Korea has taught a great lesson to all the countries in the world, especially the rogue countries of dictatorships or whatever: if you don't want to be invaded by America, get some nuclear weapons." Michael Moore

Even the U.S. Army's own historians barely mentioned Eisenhower in their book, Korea: 1951-53. Perhaps when they compiled their history they did not have access to information about the president's movements in Korea or his nuclear threats. But, historians who wrote their books later did, although many of the documents describing "Annie" were not declassified in time for some of their accounts.

A random check of different histories of the Korean War demonstrates that there was no real connection among historians regarding "Annie" and Eisenhower. T.S. Fehrenbach was an exception.

Fehrenbach was more than an historian. He was an infantry commander during the Korean War, so he recognized the value of a weapon like "Annie" as a tool in peace negotiations—and a saver of UN soldiers' lives.

Fehrenbach suggested that the Republicans and Eisenhower were unsure about how to end the Korean War. Even though Eisenhower had gone to Korea he had come back empty handed. He believed, as did many UN military leaders, that the Chinese were also looking for a way to end the fighting. But, as long as the

Russians were still involved, the negotiating impasse would stay in play. (They didn't lose interest in the Korean War until Joseph Stalin died.)

Then, Fehrenbach averred, the game changed when the U.S. Army tested its new 280mm cannon in Nevada. He wrote that UN negotiators believed that "Annie" gave them the upper hand in peace talks because "In its early years the atomic device had remained a strategic weapon, suitable for delivery against cities and industries, suitable to obliterate civilians, men, women, and children by the millions, but of no practical use on a limited battlefield-until it was fired from a field gun" (Fehrenbach 442).

Fehrenbach observed that "Atomic Annie" changed military leaders' views about the application of nuclear weapons on battlefields. He wrote that, "With an atomic cannon that could deliver tactical fires in the low-kiloton range, with great selectivity, ground warfare stood on the brink of its greatest change since the advent of firepower." He went so far as to say that "Annie" would eliminate the Communist armies' main advantage: sheer numbers.

"The atomic cannon could blow any existing fortification, even one twenty thousand yards in depth, out of existence neatly and selectively, along with the battalions that manned it. Any concentration of manpower, also, was its meat. It spelled the doom of Communist massed armies, which opposed superior firepower with numbers, and which had in 1953 no tactical nuclear weapons of their own" (Fehrenbach 442).

"Annie" Visits The Far East

Fehrenbach acknowledged, as did many of the troops in Korea who swore they saw the cannon in Korea, that the gun and nuclear shells had indeed been deployed to the Far East, although not to Korea, but someplace nearby. The most logical place, although it was not specified, would have been Japan. Then, he said, somehow word of the deployment got to the enemy. That was accompanied by rumors that UN leaders were growing impatient. They would not accept the status quo on peace talks beyond the summer of 1953.

The hopes were that the news and rumors would force the Communists to step up the peace talks. Instead, they stepped up their offensive attacks which, as any infantry or artillery trooper would attest, involved far more than the "small scale attacks" the U.S. State Department described in its history of the Korean War.

Apparently "Annie" did not intimidate the Chinese as much as President Eisenhower had hoped. At least Fehrenbach recognized the use of "Annie" as a bargaining ploy, even if UN leaders did not plan to use it on the battlefield. That set him apart from most of his historian counterparts.

Let's Use Psychology, Not Nuclear Power

Robert Leckie did not mention "Annie" in his 1963 book The War in Korea. In fact, he hardly mentioned Eisenhower. When he did, he talked about him in the context of one of the UN's biggest problems with the Communists: prisoner exchanges, a subject of ongoing concern in the peace talk process—and one of the reasons Ike was so willing to use psychological weapons and threats to get them to the bargaining table.

The UN did not want to force North Korean and Chinese prisoners of war to go back to their countries if the sides agreed on a prisoner exchange. The Communists were intransigent on that point. They insisted that all their captured prisoners be returned directly to their own countries. The impasse over the issue held up the peace talks.

Eisenhower said, "To force those people to go back to a life of terror and prosecution is something that would violate every moral standard by which America lives. Therefore, it would be unacceptable to the American code, and it cannot be done" (Leckie 157). The impasse continued.

Another historian, of more recent vintage, David Halberstam, also treated Eisenhower as a side note in his book, The Coldest Winter. He noted that Ike was elected president, and that he was a better politician than MacArthur. Halberstam said "Eisenhower was by far the more egalitarian man, a better listener and a far better compromiser. He was a general, but unlike MacArthur he never looked or sounded like a man on horseback; he seemed as natural in civvies as he did in uniform."

According to Halberstam, it was traits of that ilk that led the American voters to elect Ike as president in 1952. That and the fact that Ike promised to go to Korea, a country that most of them had never heard of, to end a war that most of them did not understand or approve of.

As Halberstam explained, "The least strident of men, Eisenhower was, the country decided, the right man to lead them into a gray, uncertain nuclear age, one in which there were not going to be total victories: he was thoughtful, strong, but not too militaristic, fair-minded and pragmatic, a man who could deal with the Russians either way, hard or soft" (Halberstam 622-23).

Eisenhower may have been all of the above, but he alone was not able to bring the Communists to the talk tables, with or without "Annie," who Halberstam never mentioned. In that respect he was in step with his fellow historians.

Stalin Finds Peace In Death—And The Combatants In Korea Reach Peace In Life

The issue of "Annie" and Ike's influence in reaching a peace pact was rendered moot in the spring of 1953. Whereas the peace negotiators were dealing directly with the Chinese and North Koreans, it was still the Russians who were the most worrisome to the UN and U.S. political leaders. Their footprints were always traceable in the Korean War, although they tried to hide their involvement.

Atomic Cannons and Nuclear Weapons

Even though Russia was playing a less obvious role in the Korean War than the Chinese and North Koreans, it was apparent to UN negotiators that it was the one country it would be dealing with most directly after—and if—a truce was reached. They were aware that the USSR had nuclear weapons, which it might have been willing to use to support the Chinese and North Koreans if the U.S. employed its own. Therefore, it was Russian leader Joseph Stalin that they were concerned with the most.

President Eisenhower and his advisors sensed that the Chinese and North Koreans were tiring of the war and leery of the introduction of nuclear weapons in Korea. That was one of the reasons he dropped hints that the U.S. was getting close to using nuclear weapons—including its new nuclear cannon. Stalin inadvertently rendered that threat moot on 5 March 1953 when he died due to complications from a stroke.

Historians agree that Stalin's death was the most significant factor in the combatants' willingness to reach a cease fire. As a result, the U.S. did not have to employ nuclear weapons and the threat of "Annie" as a factor in the war went by the board. Nevertheless, some of the troops in Korea swore that "Annie" was right there in Korea, ready to fire when needed. Maybe they were right.

Chapter 20
The Bombs Are Ready

"The atomic bomb certainly is the most powerful of all weapons, but it is conclusively powerful and effective only in the hands of the nation which controls the sky."
Lyndon B. Johnson

Allegedly, nuclear weapons were never used during the Korean War. That may not be entirely true. The Air Force may have dropped at least one A-bomb, although that is pure speculation.

Robert L. Drew, who served in the U.S. Air Force's 610 Aircraft Control and Warning Squadron in Korea, wrote the following account that appeared in The Graybeards Magazine, May-June 2012, p. 54. He suggested that the Air Force may have dropped an atomic bomb in Korea at some point. As he remembered it in his firsthand account, SINANJU HA V A NO:

"When the peace talks started, the war continued, but at a slightly scaled back pace. The peace talks started at a place called Sinanju. After a short while the peace talks were moved to Panmunjom. This is another story that I know never made it into print.

"I have read many accounts of the peace talks process, but l have never heard the name Sinanju mentioned as a peace talks site. It is as if the place doesn't even exist. But, I was there. I saw and heard with my own eyes and ears what happened.

"I am reminded of what the famous Spanish philosopher Santana said: "History is written by liars who were not there."**

**We tried to verify the fact that George Santayana actually wrote that. The closest we came was a quote from Winston Churchill, who may or may not have said, "History is written by the victors; history is full of liars." There is no direct evidence he ever said that. So, we will say the statement is attributed to Winston Churchill, but of unknown origin.

"One night, while I was on duty, we saw the biggest blip that we had ever seen on a radar scope when it appeared at the lower edge of our scopes at about 200 degrees azimuth. The track was moving north at about 350 mph. ADCC [Air Direction Control Command] asked us if we had the target. We said, 'My God, how could we miss it! This is the biggest blip ever registered on this scope.'

"ADCC then said, 'Scrub everything off your board and call in the position of this track every sweep (every 30 seconds) and don't lose it.'

"We replied, 'We won't lose it, but we still have unidentified aircraft left on our board.'

"ADCC said, 'We don't care. Scrub 'em off and follow this target.'

"This unheard of order created intense curiosity, of course, as to what kind of aircraft we were tracking. As it continued up the middle of the peninsula, we started speculating about what we were seeing. ADCC wouldn't tell us anything about it. It was top secret.

"When the object came into the range of the height finder, we found it was flying at about 35,000 feet and had just started losing altitude. It was headed for Sinanju. As it got closer to Sinanju, it lost a lot of altitude and leveled out as if in a bombing glide. It orbited over Sinanju for a few minutes, then headed east and started gaining altitude. In a few minutes it had gone off the eastern edge of our scope, probably headed back to the states.

"We all decided that this thing had to be the new B-36, our new strategic bomber, which had just become operational. The thing was so big and heavy there were only 4 or 5 places in the world where it could take off and land. Okinawa was one of these places, and it had come from that direction. The speed and altitude checked out for a B-36, and there was only one airplane in the world which could leave a signature that big on a radar scope.

"Our controller knew we were all dying to know what was going on and had put ADCC on speaker so we could all hear. Our controller kept pestering the controller at ADCC for information. Finally, the controller said, 'All I can tell you is that Sinanju HA V A NO.' This was Japanese slang for 'It doesn't exist,' or, 'it no longer exists.' This was all we needed to hear. It confirmed everything we knew about this flight.

"The B-36, the world's largest aircraft, was designed to fly to any point on the globe and back without refueling. I don't remember what the bomb load was, but it had to be phenomenal. It was our atomic bomber at the beginning of the Cold War.

"It had 6 huge reciprocating engines and 4 jet engines. It still had to have JATO rockets (Jet Assisted Take Off) to get off the ground. You could hear it well beyond visible range. All the reciprocating engines were synchronized so that matching cylinders in each respective engine fired at the same time. It was an unforgettable, awesome sound.

"The B-36 was in service only a few years because in-flight refueling was being perfected and the B-36 was too expensive to operate. The all-jet B-47 was operational for a short while, then the old B-52 workhorse took over and is so good that it is still our primary heavy bomber 50 years later.

"A day or two later we learned the true story of Sinanju. The ten-square-mile area of the peace talks was supposed to have no weapons of any kind inside the perimeter. The North Koreans and Chinese wouldn't even permit our UN officers to carry side arms. Yet, all the North Koreans and Chinese had side arms and armed guards all over the place, with international photographers taking their picture. They even had artillery and tanks inside the area. It was some kind of psychological advantage the enemies thought they were pulling off to show the world how tough they were. Orientals are big on psychological warfare.

"We told the enemies to move their artillery and tanks. When they flatly refused, we just obliterated Sinanju, along with anyone and anything inside the area. A few days later the peace talks were moved to Panmunjom, because 'Sinanju HA V A NO!'

"The enemies accused us of bombing the peace talks area, but we denied it. Why, l don't know."

Did Drew Draw The Wrong Conclusion?

Predictably, some Korean War veterans took Drew to task for his story, with good reason. One in particular, Bob Hall, refuted a lot of Drew's allegations in his letter:

"A letter entitled 'Sinanju HA V A NO' appeared on p. 54 in the May-June issue of The Graybeards. At that time I considered it a hoax, but I have since changed my opinion. I have pursued considerable research on the matter, including Clay Blair's monumental tome on the Korean War, 'The Forgotten War.' It is now obvious that the inaccuracies are the result of confusion and probably an overactive imagination.

"The claim that the peace talks began at Sinanju is not true. The writer has confused Sinanju with Kaesong, where my sources agree that they got underway on July 10, 1951. Sinanju is located on the Chongchon River, deep in North Korea, and not far south of the Yalu, a place that would be an unlikely venue for such delicate negotiations.

"The violations of the neutral atmosphere that should have prevailed and which the writer attributes to Sinanju actually occurred at Kaesong. The communists attempted to extract every possible propaganda advantage from the location.

"Incidentally, the UN was guilty of a few minor violations as well. But Far East commander General Ridgway was incensed and refused to allow the talks to continue at Kaesong. He even went so far as to defy the Joint Chiefs of Staff. But he prevailed, and the talks eventually resumed at Panmunjom on October 25.

"Kaesong today is in North Korea and Panmunjom is located in the DMZ. My wife and I were there a few years ago during a Revisit trip.

"The writer also presented a scenario involving a ghostly aerial attack by a plane which he assumed was a B-36. Of course, there is no record of such an event and I cannot categorically deny such an attack occurred. But it seems highly unlikely and is probably pure speculation that a single plane obliterated Sinanju. It is more likely that a flight of B-29 Superfortresses did inflict a devastating attack on the area.

"Sinanju occupies a strategic location that included five bridges and extensive railroad marshaling yards. USAF planes repeatedly bombed the area to interdict the flow of supplies from China and Russia headed for the battle front.

"On one occasion, 300 of these planes bombed a city on the Yalu and north of Sinanju. Allied (UN) planes bombed western North Korea throughout the three years of war, but the North Koreans (like the North Vietnamese in that war) often repaired the damage in record time."

So, did Drew's story have any truth to it? Maye, maybe not. U.S. military history is rife with operations that occurred but which were never revealed to the public, at least not until accounts of them were declassified.

The scenario Drew wrote about could have taken place. Or, maybe not. If it did, it sounds inconceivable that the North Koreans, Chinese, and Russians would keep it quiet. On the other hand, maybe they had their reasons for doing so. The mystery of the Sinanju bombing raids is still unsolved—as is the mystery of whether nuclear weapons were actually used at any time during the Korean War.

Chapter 21
What May Be The Real Story

"The United States strongly seeks a lasting agreement for the discontinuance of nuclear weapons tests. We believe that this would be an important step toward reduction of international tensions and would open the way to further agreement on substantial measures of disarmament." Dwight D. Eisenhower

Drew may very well have gotten the story right, but mixed up the facts. There are two incidents that he might have conflated. One involved an airborne unit. The other centered around a large bomb, but not of the nuclear variety. Both involved Kaesong, a strange choice for peace talks—at least for the UN.

Both sides wanted to conduct the early peace talks at a neutral site. For some reason, UN negotiators agreed to Kaesong, which was well behind enemy lines. That gave the Communists a definite home field advantage. As Blair described it:

"Right off, the UN team realized it had been a mistake to agree to meet behind enemy lines at Kaesong. Although Ridgway had insisted that Kaesong be declared a 'neutral zone,' it was hardly that. The place teemed with armed CCF and NKPA soldiers. The UN members coming by road through Communist lines had to fly white flags and go through CCF checkpoints; those arriving by helicopters were met and escorted to the meetings by armed NKPA soldiers. The Western press was barred from Kaesong; the Communist press had free access. The Communist propaganda goal was clearly to create the impression that the UN had come to Kaesong, hat in hand, to surrender or to sue for peace" (Blair, 941).

Those conditions may have led Drew to infer that the U.S. Air Force was assigned to obliterate Kaesong to force a decision to relocate the talks, and he confused it with Sinanju. There was a lot going on in the area.

As Hall suggested, Drew may have confused Sinanju, a seaport on the east coast of Korea, which is actually fifty plus miles north of the North Korean capital of Pyongyang, with Kaesong, which at least one U.S. Air Force aircraft attacked inadvertently on 10 September 1951. And what Drew may have thought was the 82nd Airborne Division was actually the 187th Airborne Regiment, aka the "Rakkasans." That is clear from a story by Col Robert I. ("Bob") Channon, U.S. Army (ret), who served with the "Rakkasans" in Korea:

"While still with the 3rd Ranger Company the Korean ceasefire planning had begun between the two sides. The initial planning location was to be at Kaesong, North Korea. Kaesong was twelve miles behind the enemy lines.

"The 187th Airborne RCT was given a Top Secret mission to be prepared to drop at Kaesong to save our negotiators from capture and then used as negotiating pawns.

(Ed. Note: The UN Command did not trust the CCF-NK to negotiate in good faith and feared they would imprison UN negotiators and threaten harm if their wishes were not met. Ergo, the plan to rescue them.)

"However, the 187th had recently suffered very severe casualties at Imjin, Korea. They needed Airborne-trained, combat experienced replacements immediately. The six Airborne Ranger Companies were the ideal source for these replacements—enlisted men, NCOs, and officers. It had been decided to reactivate the Ranger Companies and merge them into the 187th.

"In order to simultaneously drop the entire 187th, two major airfields were required. There was no place in Korea where two major airfields were close enough together to permit proper training.

"On the Island of Kyushu in Japan, Ashiya and Brady airfields were ideally positioned on the northwest coast to support the 187th's mission. (Ed. Note: Thus, though the 187" was the strategic reserve of Eighth Army and UN Forces, it was moved to Japan. Though this added about another hour+ flight time to the proposed [drop zones] near Kaesong, it was not considered critical.)

"So, in about June 1951, orders were issued to move the 187th to Japan. I was not sent back to 3rd Airborne Ranger Company and was assigned as Plans Officer of the 187th to coordinate the move. (Channon's Note: Your past always catches up to you! In view of my prior experience in planning and coordinating movement of the 11th Airborne Division from Japan to the U.S.A. in 1948-49, I got the detail to plan and coordinate the 187th's movement to Japan.)

"We started moving the 187th to Japan in July-August 1951. The six Airborne Ranger Companies arrived in Japan at the start of September 1951. That is how the Rangers ended up in Japan as part of the 187th Top Secret mission!

"Training for the 'rescue' mission was grueling and interrupted when North Korean POWs on the Island of Koje-do off the south coast of Korea created a revolt that required aggressive action to constrain."

So, the 187th never carried out a mission at Kaesong—or Sinanju. Neither did the 82nd Division. As a result, Drew's account of events at Sinanju may have been skewed. In any case, his "facts" predate President Eisenhower's December 1952 visit to Korea. But, his is the type of misunderstanding that is at the center of the mystery surrounding the connections among Ike, "Annie," and peace in Korea.

Tarzon, Not The "Ape Man"

There is another possibility: Drew may have confused the incident with a serious mishap over Sinanju involving an A-1 Tarzon bomb, aka VB-13, and B-29 bombers from the 19th Bomb Group. Tarzon, the largest bomb deployed operationally in the Korean War by the U.S. Air Force, was a guided bomb developed by the United States Army Air Forces during the late 1940s. It was created by integrating the guidance system of its predecessor, the Razon radio-controlled weapon, and a British Tallboy 12,000-pound (5,400 kg) bomb.

Tarzon was used for a short while in the Korean War, but its unreliability (only 6 of the 28 bombs dropped actually destroyed their targets), design flaws, and high costs led to its premature retirement in 1951 after only thirty missions. The problems included increased maintenance costs compared to conventional bombs and the need to release the weapon at a prime altitude.

More important, Tarzon could only be dropped in clear daylight to take advantage of its guidance system. That requirement exposed the B-29 and its crew to enemy fighters and made it extremely vulnerable to enemy flak.

One incident over Sinuiju vividly highlighted those flaws, and may have led Drew to conclude that a nuclear bomb had been deployed because of the similarity in names (Sinanju vs. Sinuiju, which are about 73 miles apart). A single bomb, including its control surfaces and annular wing, weighed 13,100 pounds. The bomb was so big a B-29 could not carry it inside the bomb bay. It was attached to a semi-recessed mounting, which left the bomb exposed to the airstream and created a turbulent airflow. That affected the plane's handling and made the pilot's job more difficult.

The 19th Bomb Group dropped the first Tarzon in combat on December 14, 1950. Despite its flaws, Tarzon was a very effective weapon against high-priority targets such as bridges and hydroelectric plants that "dumb bombs" hardly damaged—if they hit them. But, its efficacy came at a price, as the aforementioned incident over Sinuiju demonstrated.

The group commander's plane suffered a mechanical failure. The crew tried to jettison the Tarzon in preparation for ditching the aircraft. It exploded prematurely and blew the aircraft apart. Three weeks later there was another incident of the same type, although this time the aircraft was spared.

The Air Force conducted an investigation to find out why the Tarzons were exploding prematurely. Investigators determined that the problem was a flaw in the bomb's tail. Technicians made some modifications to resolve the issue, but they were too little too late. Production was halted and the Tarzon program was discontinued in August 1951. Tarzon may have been Drew's assumed nuclear weapon.

One Bomb: 200,000 Casualties

Drew was not alone in telling a story about a cataclysmic nuclear bomb that caused heavy casualties in Korea. There is an internet report that tells the story of an atomic bomb allegedly dropped on Teng Sha Ho Military Base in Liaoning, China that was so devastating it forced the Chinese to sue for peace. However, there is no evidence to support the writer's story, which is:

"Under pressure from the public and Congress, Harry Truman decided that he should listen to MacArthur's advice and 'apply every available mean to bring it to a swift end.' Already stationed in Guam were 10 B-29 Superfortresses and all components to a nuclear missile except for the fissile core. On April 23, an atomic bomb was prepared and sent to Kadena AFB on Okinawa, Japan.

"On the morning of April 24, 1951 Colonel Kenny Roberson

took off in the B-29 Superfortress Praetorian. On board was the Mark 4 Atomic Bomb, 'Big Boy.' Roberson steered the plane the needed 949 miles from Kadena AFB to Teng Sha Ho Military Base, in Liaoning, China. The targeted base was just outside of Dalian, a large city that has traditionally been home to many Chinese military bases.

"The effects were much more devastating than the atomic bombs at Hiroshima and Nagasaki. Big Boy killed 150,000 Chinese soldiers and military personnel and an additional 50,000 civilians. The base was the launching point for Chinese reinforcements, and was also undergoing a construction project to build an air base.

"Teng Sha Ho Military Base was 949 miles from Okinawa's Kadena AFB. It is 222 miles from Pyongyang and 558 miles from Seoul."

Evidence for such an event ever occurring is difficult to find—or nonexistent. There is anecdotal evidence, however, that UN planes did cross the Manchurian border on a semi-regular basis. Historian Jon Halliday wrote that allied planes attacked a Russian air base in China and Chinese Army headquarters at Dandong. 'The only thing which was strictly vetoed, I was informed, was attacking Soviet planes or military personnel on the ground-though one such attack did take place, apparently in 1952, on a Soviet air base in China.'"

And, Halliday noted, "Major General John G. Singlaub states that 'the [U.S.] Air Force was . . . flying regular combat missions north of the Yalu,' but he dates this only "toward the end of the war" (Halliday 155). Based on observations such as those, it is entirely possible that allied planes did attack Teng Sha Ho Military Base. It is less conceivable that they inflicted 200,000 casualties on the Chinese. That statistic would have made the nightly news in at least some of the countries involved.

But, the story is typical of the mystery surrounding the role of nuclear weapons in Korea. They were like the Loch Ness monster, Yeti, and the Abominable Snowman. They may all be figments of people's imaginations, but there may be an element of truth regarding their existence. At least the U.S. Air Force did have B-29 Superfortresses and nuclear weapons on site at Kadena—and even practiced dropping the bombs.

Significantly, Air Force pilots conducted practice runs through Korea in Operation Hudson Harbor in October 1951. They wanted to be ready should the use of nuclear weapons be approved. The U.S. Navy had them available as well.

Operation Hudson Harbor

In October 1951 the U.S. implemented Operation Hudson Harbor to send a message to the enemy regarding the use of nuclear weapons, prepare Air Force pilots for actual nuclear bombings, and scare the enemy by demonstrating how effortlessly U.S. planes could penetrate their air space and drop the bombs. One of the secondary goals was to warn the Russians to stay out of the way.

The operation involved bomb runs by B-29s flying from Okinawa to pass over pre-selected targets in North Korea. The planes carried and dropped either conventional bombs or dummy nuclear bombs. Essentially what the Air Force was doing was testing air and ground crews' abilities to assemble, test, load, and drop nuclear bombs, and making sure communications channels were sufficient for adequate control of the missions should the order ever come to employ them. It never did, and Operation Hudson Harbor ended with a whimper, not with a bang, nuclear or otherwise.

Chapter 22
Anchors And A-Bombs Aweigh

"The 20th century was a test bed for big ideas - fascism, communism, the atomic bomb." P. J. O'Rourke

As peace talks began in Korea there were rumors galore alluding to the availability of nuclear weapons in the U.S. arsenal and the military and political leaders' willingness to use them. They may have been rumors; they may have been facts. In all probability they were a combination of rumors and facts.

The troops doing the actual fighting on land and sea and in the air would seize on any facts or rumors if they believed they would get them home quickly. MacArthur had promised in 1950 that they would be home for Christmas. They weren't. By early 1951 it became obvious that their chances of getting home that year at Christmas or any other time were dim—and getting dimmer as the year progressed.

1951 became 1952, 1952 became 1953, and millions of American and UN warfighters had rotated through Korea. Service members, politicians, citizens…everyone was tired of the war. It was no surprise, then, that rumors about the use of nuclear weapons became more widespread, and "sightings" of them were reported by troops in the field and sailors at sea. At least two U.S. Navy aircraft carriers, Lake Champlain (CV-39) and Valley Forge (CV-45), were equipped with nuclear weapons and crews trained to deliver them.

That Ain't Food In Those Coolers

Tom Moore, a U.S. Navy Korean War veteran, revealed that the Navy was prepared to distribute nuclear bombs in Korea by mid-1953. It stored atomic weapons aboard the aircraft carrier USS Champlain and carried planes assigned to drop them. One of the pilots assigned to do the job told Moore the story. As he wrote in the March/April 2010 edition of The Graybeards (p.22):

"In 1953, 7th Fleet Commander Vice Admiral J. J. ("Jocko") Clark obtained permission from Commander-in-Chief, United Nations Command, General Mark W. Clark, to arm his 7th Fleet fast carriers with nuclear bombs. Soon, in Task Force 77, the USS Lake Champlain (CVA-39) had refrigeration men running refrigeration lines in storerooms below decks.

"Then, special equipment and lifts were installed in these storerooms. The refrigeration-men were kept away from the crew. You could spot them by their radiation badges. When everything was ready for the nuclear bombs, Lake Champlain went to Sasebo, Japan. The A-bombs were brought out on the dock on rubber-tired wagons, with huge suspensions. The dock was lined with U.S. Marine guards. Once loaded, the ship was off, back to the war zone.

"The A-bombs were stored in the refrigerated storerooms in sections. If need be, they would be assembled in a storeroom near an elevator. Lake Champlain had the SCB-27 modernization. (See Appendix D for a description of SCB-27.) Among other things, the flight deck, elevators, catapults, and arresting gear were reinforced to support aircraft weighing up to 52,000 pounds, mainly the AJ Savage bomber. (See a description of the AJ Savage bomber in Appendix E.)

"The AJ Savage (the first naval combat aircraft designed to carry an atomic bomb) had worked with the ship before, with sand bombs, for proper weight. To deliver the nuclear bombs, the ship could use AJ aircraft from VC-6, Atsugi, Japan. With special new ordnance equipment, atomic bomb releases, etc., for VA-45 Skyraiders, the AJs could be used.

"When the 7th Fleet was ready to deliver the nuclear bombs, the word was leaked to the enemy. Later, Vice Admiral C. Turner Joy, the UN delegate at Panmunjom, stated that the Communists got busy and speeded up the cease-fire talks. That led to the cease fire signing, when they learned our fleet had the A-bomb, which saved many UN lives.

"After the cease fire, Lake Champlain carried the nuclear bombs, 'just in case.' The ship off-loaded the bombs in October 1953, at Sasebo, Japan."

Moore added a sidebar: LT (jg) George "Gus" Kinnear II, later a four-star admiral, of

Squadron VA-45, was to fly his Skyraider AD aircraft from the deck of Lake Champlain to make the first atomic bomb attack, scheduled for 1 August 1953. But, the fighting ended four days before that.

Valley Forge, Nuclear Weapons, and the VF54 Fighter Squadron

On October 11, 2010 Jack L. Hatchitt sent a request to USN Commander (ret.) Glenn War, who had served aboard Valley Forge, to clear up some theories about the presence aboard or use of nuclear weapons by VF-54 pilots. Hatchitt wrote:

"LTJG Alvis and I commandeered a two-ton truck and went to Camp Elliot for a supply run to the 'Nuclear Facility Weapons Depot' located within the camp. We first had to go to the Miramar Armory and get checked out with Colt .45 side arms. Then we left to go to Camp Elliot. There were armed guards with machine guns, razor wire, outlook towers with machine guns, etc. I had never seen so much security; it was everywhere.

"Not only did we have to show our credentials to get into

Camp Elliot, but we had to show 'Top Secret' credentials to get into the nuclear facility. Once inside we loaded the truck with a covered, crated object that was quite long. As we left, we were stopped at the gate by guards and ordered out of the truck. Somewhere along the line our authorization for this object was not in order.

"We were placed in a holding area until it was straightened out about three or four hours later. I thought the verification had to come from someone in a higher authority than our skipper, LCDR (Lieutenant Commander). Suerstedt. Perhaps it was CINPAC (Commander in Chief Pacific).

"We finally returned to NAS Miramar later in the day. The crated object was turned over to Ordinance, then put in a corner of the hangar, where a 24/7 guard was stationed with it. I had to sign documents of nondisclosure with severe penalties in a breach of contract. I could have gone to jail for several years.

"I could not talk to anyone about our trip or reveal what we picked up—and I mean anyone. However, I believe what we picked up was 'equipment' for arming our AD aircraft with nuclear weapons. It is just my theory.

"When we deployed to NAS El Centro our pilots practiced 'loft bombing.' This was an exercise that was to be used in launching a nuclear weapon. The pilot flew level, then pulled up going skyward, lofted the bomb, flew out on his back, righted the plane, kicked in the alcohol injection system for a faster getaway, and then prayed.

"One of our pilots was killed during this exercise, because after he performed the 'loft' he never pulled out and plunged into the desert floor.

"While we were aboard the USS Valley Forge our aircraft were being fitted with nuclear capabilities. The ordinance men were doing the work behind shrouded enclosures of our aircraft so no one could see what they were doing. Ensigns Stoddard and

Serfus were assigned the task of keeping curious people away from that area. They also carried side arm weapons.

"Alcohol injection systems were also installed on our aircraft for a quicker means of exiting the target area. The pilots would 'loft bomb,' come out on their backs, kick in the alcohol injections, right the aircraft, and go like a bat out of hell to a safer area.

"I learned later on that General MacArthur had ordered nine nuclear bombs to be used in Korea and possibly Japan. General Curtis LeMay of the Strategic Air Command was outraged that General MacArthur passed him by to get the bombs."

Hatchitt also referenced the aforementioned California bomber crash in his account, since it was connected to VF-54.

"Nine B-17's took off. One crashed. There were no problems with the bomb aboard. However, the other eight went on to Guam, from where they were to be delivered to VF-54. We were then designated a Nuclear Squadron, which accounted for the loft bombing techniques.

"President Truman stopped General MacArthur's proposal to use the nuclear weapons in Korea and possibly in China. I wonder if his wasn't why President Truman fired General MacArthur."

Hatchitt did not provide a timeline of his activity at Camp Elliott, but it had to be between November 1952 and June 1953. LCDR Henry Suerstedt did not take command of Valley Forge until March 1952, at the same time it moved from NAS North Island to the Master Jet Air Station at Miramar.

Ward responded the next day:

"All this stuff occurred after I left the squadron. You and LTJG Alvis undoubtedly picked up a 'Shape,' a dummy bomb with weight and dimensions of the real thing, designed to give the crews experience with loading, handling, etc."

That prompted an addendum from Hatchitt to Ward that highlighted the seriousness of the situation:

"Forgot to add that two or three days later, LTJG Alvis and I were called into the Skipper's (LCDR Suerstedt) office. When we arrived there were three men dressed in suits. They were introduced as FBI agents. They wanted to question LTJG Alvis and me about our trip to the 'Nuclear Facility' at Camp Elliot.

"For some reason or another we were never logged out of there and they wanted to know what we did for the 36 hours while we were there. They threatened Leavenworth Penitentiary, all kinds of penalties, etc. We told them we had logged out and if they would check the squadron log-in sheet for that day it would show we returned about five hours later.

"After some more questioning we were finally told to go, but they said they may be back for more information. I thought sure we were going to be shot at sunrise.

"Due to all this I had to attend ('RADSAFE'), Radiation Atomic Defense Safety School at North Island Naval Air Station, San Diego for one intensive week. I learned that Lithium-6 is used to increase the detonation of the nuclear reaction in the bomb. I still have my credentials authorizing me as being qualified for this event."

Chapter 23
How Nuclear Bombs Saved Lives

"The Air Force and the Navy carriers may have kept us from losing the war, but they were denied the opportunity of influencing the outcome decisively in our favor. They gained complete mastery of the skies, gave magnificent support to the infantry, destroyed every worthwhile target in North Korea and took a costly toll of enemy personnel and supplies. But...our air power could not keep a steady stream of enemy supplies and reinforcements from reaching the battle line. Air could not "isolate" the front. This made it a footslogger's war.

To have pushed that war to a conclusion in the mud and mountains of Korea would have required more trained divisions, more supporting air and naval forces, would have incurred staggering casualties and could not have been attempted with any hope of success unless we had lifted the self-imposed tactical restrictions which gave the enemy a sanctuary north of the Yalu. I believe, however, that we could have obtained better truce terms, shortened the war and saved lives, if we had got tougher faster." General Mark Wayne Clark

Moore's story re Lake Champlain fascinated U.S. Army veteran Andrew Antippas, who tied it to the rumors surrounding "Annie" in a letter in the May-June 2010 Graybeards, p. 60, and inadvertently reignited arguments about how many service members' lives the use of the atomic bombs in Japan in 1945 had saved. There were "experts" who suggested that as many as one million military and civilian lives were saved. They would otherwise have died in the final assaults on Okinawa and the Japanese mainland if the bombs had not forced the Japanese to surrender.

General Clark, who was the commander of UN forces in Korea from May 12, 1952, to October 7, 1953, responded to a gentleman who asked him if he believed that dropping the atomic bombs on Japan was the right thing to do: "Yes, I believe we should have dropped the bomb as we did. Otherwise, we would have had battle casualties in the hundreds of thousands of our fine men on landing on the shores of Japan, where every person would have had a gun. It ended the war and saved lives."

That same scenario may have been played out in Korea if nuclear weapons were not used. Fortunately, it never came to that in Korea, although Clark might not have hesitated to use them if necessary. He did push bombing boundaries in the later stages of the war.

Early in 1953 Clark ordered U.S. forces to bomb the North Korean capital of Pyongyang, hydroelectric plants along the Yalu River, which were dangerously close to China, and other targets that had been declared off limits earlier. The raids all but decimated Pyongyang. The air attacks on the hydroelectric plants turned off electric power in North Korea for two weeks.

The bombing was so severe that even the British and French protested against their intensity. Clark did not relent. His goal was to get the Communists to agree to an armistice and end the war, and a few friendly protests were not going to stop him. There can be no doubt that Communist leaders kept in the backs of their minds the thought that Clark would not hesitate to use nuclear weapons to accomplish his aim.

Why not? There were nuclear weapons on Okinawa and aboard USS Lake Champlain. "Annie" was rumored to be ready for service in Korea. The use of nuclear weapons was the next logical step.

Antippas wrote:

"I was fascinated by [the] report in the March-April 2010 issue of The Graybeards

(p. 22) on the assignment of USS Lake Champlain (CVA 39) to carry atomic weapons for possible use in the Korean War in the spring of 1953. At that moment, I was an infantry squad leader in the 45th Division on the eastern front.

"I recollect the scuttlebutt at the time about the leaking of the word that atomic weapons were in the Korean Theater and that President Eisenhower was frustrated at the delay in completing the armistice agreement. The issue was complicated by Korean President Syngman Rhee's unilateral release of the 25,000 [North] Korean POWs under ROK control in mid-June 1953 in an effort to sabotage the armistice. (My battalion was initially alerted to go out and hunt down the escapees.)

"The net result was the angering of everyone on both sides and the decision of the communists to give the ROKs a bloody nose. Hence, we experienced the last bloody campaign of June and July with the ROKs taking the bulk of the punishment.

"With the 280mm atomic cannon having been test fired in May 1953 in Nevada, we assumed that the atomic weapons speculated about was 'Atomic Annie.' In any event, we all knew that if the armistice talks failed once again, we well might have to attack through the reputed 20-mile deep defensive zone that the communists had constructed along the whole front.

"Knowing what we know now about radiation effects and the communist defensive preparations, attacking would have been problematic for both sides."

Atomic Cannons and Nuclear Weapons

It's Not Murphy's Law, But It's Murphy's Explanation

U.S. negotiators trying to end the Korean War in the early stages were not shy about letting the enemy know they had atomic weapons at their disposal and would not hesitate to use them. They had another secret, which they kept to themselves: the atomic cannon did not really exist in tangible form, at least not at that point. Or did it?

There were military personnel who knew circa 1951 that "Annie" was in the works, but that it might never be used as anything but a threat. According to retired U.S. Marine Corps Major James L. Murphy, who graduated from the Army Engineer Career Course at Ft. Belvoir, VA as a qualified Nuclear Weapons Employment Officer, the Army, together with other agencies, developed an artillery piece ("cannon") with the capability of delivering an atomic ("nuclear") shell.

Murphy observed that "The weapon used the same design as in the 'Little Boy' atomic bomb dropped on Japan; a ring of nuclear material with an insert, that, when triggered, slipped into this ring and became "critical," that is, "explosive."

Murphy added that the system was tested in New Mexico and performed as intended. But, he declared, "The debate as to whether this 'cannon' and related atomic shells were ever deployed in Korea is questionable as the political climate would have precluded the use of any atomic devices, this in spite of some who thought General MacArthur had advocated such bombing of China."

And, he pointed out, "To deploy and use such devices would have required authority from the highest level--presidential, in all probability." That was just one of the problems with the atomic cannon.

We Don't Care What We Didn't See: We Saw It

Surprisingly, large numbers of U.S. military personnel inside and outside Korea during the war were convinced they had seen or heard the atomic cannon, or evidence of its existence, as early as 1951. According to one former Air Force member, Dick Larrowe, even the North Koreans got a glimpse of it.

"I never saw the atomic cannon in Korea," he acknowledged. "However, I did see it at Fort Sill, Oklahoma in the 1980s while I was visiting my son who was in the Air Force and stationed at Shepherd Air Force Base."

Larrowe explained that a display at the Fort Sill museum noted that the gun was never fired at the enemy, but President Eisenhower arranged for North Korean generals to look at one. As Larrowe speculated, "This was a deciding factor for North Korea in agreeing to a cease fire in 1953." And, he concluded, "It probably explains why North Korea acts the way it does now that it has its own atomic bomb."

Most of the people who saw or heard the gun were members of the U.S. Army. There were exceptions. Ralph Delaney, a Marine with B Co., 1st Bn., 1st Regt., heard it.

"On my second night in Korea I was trying to sleep over the noise. About 9 p.m., a cannon went off just over the hill behind me," he said. "My cot, which was on a dirt floor, lifted off the ground. In fact, the entire squad tent came off the dirt. I thought that it was all over for me--and I had just gotten to Korea."

Delaney learned from other Marines the next day that the gun he heard was the "atomic cannon." "It fired for a while," he concluded. "But I never saw it." That was the case with a lot of Korean War veterans. They "heard" the gun, but never saw it.

Sgt. Wayne A. Schild heard the 280 in 1953. He was stationed near Munsan-ni with the 84th ECB (Engineer Combat Battalion) before the armistice was signed.

"We were building the project X-Ray Bridge over the Imjin River," he recalled. "Every so often a loud, thundering noise would awaken us as a massive projectile went overhead into North Korea.

We were told that it was one of the 280mm cannons that were positioned over a couple of nearby hills."

Schild saw one of the 280s later at the 1957 inauguration parade for "General Ike," as some of the troops so fondly called him--almost four years after he heard it. That was the story of the atomic cannon in the 1950s in Korea: some people heard it, some people saw it. They seldom saw and heard it at the same time.

Chapter 24
Was "Annie" Twins?

"I think there are many times when it would be most efficient to use nuclear weapons. However, the public opinion in this country and throughout the world throw up their hands in horror when you mention nuclear weapons, just because of the propaganda that's been fed to them." Curtis LeMay

John Sonley was among the Americans who saw the atomic cannon. "I remember the atomic cannon very well," he said. "There were two of them in the rear. I learned about the second one from a friend about twenty years after I was home. He said he was stationed close to where they were stored in a rear area.

"It was about April in 1951 when my company encountered the first cannon, which had been brought up to our position. We were on the move all day when we were ordered to dig in at the top of a small hill for the night. The night was pitch black. We could not see our hands in front of our faces. Going up required us to hold on to the cartridge belt or hand of the man in front of us, so we would not step off the path.

"As we worked our way up, I had to really strain my eyes in order to keep close to my leader. Suddenly I went totally blind; I could not see a thing. I lost contact with the man in front of me, and had to call out softly and ask him to step back and help me. The man in back of me told me to step aside and said he would help me up the trail, since he could see. All I could think about was that I was going to be blind for life, and this was scaring me.

Atomic Cannons and Nuclear Weapons

"As we made the crest of the hill, searchlights lit the sky as they bounced their beams off the low clouds to the ground. That was when my eyesight returned. We had dug in for the night, when we heard a racket at the side of the hill and on a road just below us. All I could think of was why we had come up the side of the hill instead of using the road.

"We went to the edge, where we were able to make out a large cannon on a carriage being maneuvered around a curve in the road by one tracked vehicle in front and one in the rear. It was of a light color and looked to me like it was something from a battleship. There was a crew working around the cannon, so we went back to our positions.

"Some time later, maybe an hour or so, without any warning to us, this cannon was fired to the north of our position. There was a tremendous white light from the muzzle, followed by a deafening sound. Then, all was still. Every one of us was afraid to move, as we thought for a split second it was 'incoming.'

"We then heard the tractors moving the cannon back around the curve as we waited for the sound of the shell landing someplace. We never did. Our C.O., Captain Lincoln, said the cannon could fire a shell over twenty miles or more, and we wouldn't hear it land. He was right; we never did that night.

"We were ordered to remain in our position for a few days, but we were never told why. The next night, the cannon was brought almost to the front of our position, but just enough so the muzzle was sticking out around the curve of the road. We were ready for the cannon to shoot this time, so we hit the ground. After the cannon fired, it was again moved around the back of our position. Still, no one told us what was going on with this cannon. If the Commanding Officer knew, he never said a word.

"Annie" Dies

"The third night, it was the same thing again, with one

exception: the ending was not what anyone would have expected. Again, we hit the ground as it fired. Then, I think every artillery piece the Chinese possessed had been brought to the front, as the incoming shells landed on and around the atomic cannon.

"Believe it or not, not one incoming round landed on our position. At first light that morning, everyone ran to the edge of the hill to see what had happened below us. We saw the cannon lying on its side, off the road. The tractors were demolished and in pieces.

"After we had a C-Ration breakfast, we were ordered to vacate the hill and move forward. Much of the talk was that the cannon had fired a small atomic shell north as a test and we were going to be fighting an 'Atomic War.' But, a word was never said to the troops as to what it was all about. Not even the officers would talk to us about what took place.

"If the atomic cannon was ever used again, we never knew. My friend said that as far as he knew, the two were never moved out again while he was there."

The fact that Sonley placed "Annie" in Korea in 1951 suggests that he may have seen a 240mm cannon. It is highly unlikely that the Army would have a cannon in the field two years before it was actually tested. There were, however, service members who professed to see "Annie" in 1953, the year it was field tested. They included Robert Palmrose and John Marin Campbell.

It Was There; We Saw It

Let's start with Palmrose:

"I served with the 40th Division Artillery, HQ & HQ Battery, Message Center, August 1952 to August 1953. During this period we moved 4-5 times, starting from Kumwha Valley. Later, we went back there after other moves in areas of the north and northeast.

"On one occasion I was with a motor messenger in a jeep when we came to an area with a very large 'piece,' not the usual 105 or 155 cannon we were used to. We learned that two tractors pulled the piece, and that it took about a day to set it up for firing. We also found out that there were three of the guns in Korea, one in each Corps area.

"The cannon could fire shells eighteen miles with pinpoint accuracy. They were firing once or twice a day, but with 280mm regular powder, not atomic at this time. The crew told us all this."

Campbell, who served with the U.S. Air Force's Air Resupply and Communications Service (ARCS), tells a similar story. Despite its innocuous sounding name, ACRS was anything but a supply and communications service unit. It was a secretive unit responsible for carrying out psychological warfare missions. Its planes were unmarked and its members were not always aware of what they were actually doing.

Art Snyder, an ARCS pilot, explained that "The Wing was formed at Mountain Home AFB, Idaho in February 1951 as a CIA Wing to fly psychological warfare missions--in violation of the Geneva Convention, we were told. As such, members were war criminals...Our B-29s were unarmed and painted black with no USAF markings... We were told what to do on our missions, but we were never told why we did it...One of the problems with being in the 581st for career officers was that it was a dead end assignment. No officers from the wing were ever given a command assignment."

Campbell's situation was not quite that dire. He was out and about with troops in the open as an observer. It was during one routine assignment in 1953 that he saw "Annie," which no one appeared to be trying to hide. Campbell explained:

"One day, as Allied observers, we watched a platoon of ROK infantrymen wearing freshly cut foliage on their helmets practice attacking another ROK platoon in a wooded canyon. Another day, with a borrowed military police riot gun, we went looking for spectral pheasants that Ross had seen the week before in some abandoned rice paddies. Our third day out, on a heavily traveled dirt road behind the forward ridges, we found an exceptionally big, long- barreled field gun that had just been pulled down from the Main Line of Resistance (MLR) by an Army tractor and a dozen U.S. artillerymen.

"Neither Bill nor I had seen another like it, nor did we know what it was meant for until an artillery captain explained that it was an atomic cannon. Its low-yield shells were designed to wreak havoc among enemy companies and battalions. He and his men had been trained in its operation, which was the reason they were there. But, he said, orders had come down at the last minute to scrub the mission and get the cannon back to where it had come from."

If the cannon did not fire, would it have been destroyed as Sonley claimed he had seen? The answer falls into the "who knows" category, as does much of the mystery surrounding "Annie's" role in the Korean War.

No Need For A Eulogy For "Annie"

It is understandable that the U.S. Army would never admit to having one of its atomic weapons demolished by Chinese artillery, as Sonley reported. Then again, the Army swore that none of them were in Korea during the war. So, if it wasn't there, it couldn't have been demolished. Therefore, in a classic case of circular reasoning, Sonley and his comrades could not have seen a cannon that was not there.

The mystery deepens. And, there may be reason to doubt an eyewitness if there is only one (or several in the above case). But Sonley and Campbell were not the only eyewitnesses to what they may or may not have seen.

U.S. Army tank commander Lindy Constantino and his crew

had an up-close-and-personal encounter with what looked to them like a family of "Annies" in 1953. Could they have been wrong too?

Chapter 25
"I Heard From a Friend Who Heard From A Friend...."

"When you see something that is technically sweet, you go ahead and do it and you argue about what to do about it only after you have had your technical success. That is the way it was with the atomic bomb." J. Robert Oppenheimer

Certainly, if the atomic cannons were moved, someone would have noticed. They were hard to miss, as Lindy Constantino suggested.

"On a work shift with the 14th Combat Engineers, Company B, somewhere between Yongdong-po and the road to the Injin River, in either the summer of 1952 or the spring-summer of 1953, we came across the atomic cannon," he said. "We were all totally amazed at the sight." Like so many of his contemporaries, he always wondered where that cannon ended up.

No "Lalley Gagging" Please

Tony Lalley was among the soldiers who knew the atomic cannon existed in Korea--even though the official history of the gun states that they were stationed only in West Germany and Okinawa. (See the "official history" at http://www.theatomiccannon.com/history.)

Lalley was sure that he had seen evidence of atomic guns in Korea. In fact, he almost ran over their commanding officer with his tank. As he explained:

"There was at least one artillery unit in Korea with the

Atomic Cannons and Nuclear Weapons

capability of firing atomic warheads. Not that I know this personally. Rather, I know just from being told such by reliable sources and several personally viewed clues.

"I was in Korea from December 1952 through November 1953. While there, I was always an 'official' member of the 4th Platoon, Company 8, 89th Tank Battalion, 25th Infantry Division. I describe it this way because I or my entire platoon was often assigned TDY (temporary duty) to other 25th [Division] units.

A Cease Fire By Any Other Name...Just Stop Shooting At Me

"This event took place after the 'cessation of hostilities.' (What a joke that expression is! That's why I like to use it instead of other nomenclature like 'cease fire' and 'truce'.) My best guess would be October 1953. By then, Company B would have completed our transition to the new Patton M47 tanks and our training completed. (A great tank!)

"I didn't know our exact location along the line below the DMZ, but an educated guess would be somewhat east of the Kaesong/Panmunjom/Munsan-ni/Freedom Road corridor; the Central Front.

"Our Company B HQ and camp was located within a large encampment containing the entire battalion. We were located about two miles south of the DMZ. The entire 8th Army, along with the Republic of Korea Army, was training and preparing for an expected invasion from the communist forces in the north.

"The defensive strategy as described to us was for minimal ground forces to 'hold the DMZ line' with a delaying action while the entire remaining UN forces moved to take up positions manning a previously constructed defensive line further south. (Names like Kansas and Wyoming as line names seems to ring a bell.)

"Once those movements were completed, the 'holding forces' would be given the signal to HA ("haul ass") to rejoin their division positions further south. The main counter-strategy to the invasion was to be a massive artillery and aerial bombardment. The air assets would be coming from not only 'in country,' but Japan and other Pacific locations.

"Both 'A' and 'B' Companies of the 89th were assigned to be part of the forces holding the current line. From this point forward, I will speak generally for Company B but with more specificity for my 4th Platoon.

"The signal for an impending attack was a very loud, wailing siren. This occurred whenever a threat was observed or when the high command decided upon a surprise drill! The bottom line? We went out at least once a week.

Don't Pull Over Even If You Hear The Sirens

"By then, I was the 4th Platoon Sergeant and as such commanded the '44' tank. Our Platoon Leader was a U.S. Military Academy grad of 1952 who commanded '46' tank. '42', '43', and '45' tanks completed the platoon.

"The company's tank park was located with the company maintenance platoon and was about 150 yards up the road from Company B's bivouac area. When the sirens went off, everyone raced to join their crew on the tank. All assignments were for the platoon. Therefore, when a platoon was ready to pull out, the lieutenant radioed such and we pulled out with '46' in the lead and '44' in the rear of the column.

"We traveled about two miles at pretty high speed along the MSR (main supply route), and turned off onto a secondary road which headed for the locations of elements of the 35th Infantry Regiment. At pre-designated spots for each tank, our squads of infantry were waiting to board our tanks. We then continued about another mile to our assigned and prepared defensive position on the DMZ.

"During the route, at the spot where we exited the MSR onto

the road to the 35th, and on the right immediately after the turn, there was an 8th Army Artillery unit in position. I don't remember its unit number, but their pieces were those big 'Mommas.' I think they were 280mm cannons and they could fire both conventional and nuclear warheads. The only shells larger were on naval ships. That data is not a guess, but a fact.

Is "Annie" Cloning Herself?

"I believe there were three of those things in there. It was hard to see, but leaning up from my TC hatch enabled me to see a lot more and I could see all of one gun. The camp was very fenced and secure. I can't say whether or not nuclear tipped ammo was actually there, but all things considered, and knowing that the battery was an integral part of the initial defense, having them made sense.

"When the 'All Clear' was received over the net, the return routine was somewhat modified. The route remained the same, but in reverse. The exit sequence and column were predetermined. A minor number of vehicles and troops left their tactical positions, came to the exit point from the DMZ area, and turned down the road heading for the MSR. They, in turn, were followed by the tank platoons with their infantry.

"Our 4th platoon, with mounted infantry, was assigned as rear guard for that sector. When our platoon leader signaled his 'All Clear,' he proceeded to the exit road and departed, followed by our other tanks. I, with '44' tank, crew and infantry squad, as ordered, moved to the corner, noted by radio that I arrived there, stopped and established that there were no stragglers in view, radioed that info to the lieutenant and, at his okay, turned and made our exit.

"This part of our procedure generated a large gap of distance/time between '44' tank and the rest (19 tanks and other vehicles) of the overall column heading for the MSR. This anomaly will contribute to what follows.

Where The Heck Did That Jeep Come From?

"We started down the hill at a measured pace with our infantry hanging on. When we arrived at the drop-off place, we waved to our "Cacti" (35th Regiment) grunts, got the 'All Clear' from their platoon leader, and got out of there. Then, we could pick up speed to catch up with the column.

"My driver, Cpl. Ray Bratcher, a former D6 Dozer operator from Bowling Green, KY, could handle that M47 as well as Mario Andretti handled his race car. We were way behind the column, but he was capable of closing the gap. We were moving along, sometimes downhill, at a rapid clip.

"The interesting thing about these Patton tanks was that they were relatively quiet, especially when they weren't pulling a hill or a load. Frankly, all we would hear in such a situation was the 'clackety, clack' of the tracks.

"Down the road and around the bend we went. As the entrance to the MSR appeared ahead, a jeep pulled out of the artillery unit and turned left for the MSR. The driver never even looked our way. Now, imagine: the alert is over, the long dusty column has gone by, and the jeep and its passengers could get going. Surprise!!!

"It was a command jeep with its back seat full of radios and cat whip antennas. A fancy painted spare tire cover sported a unit designation and name--which included the nuclear symbol.

It was the commanding officer of the artillery battery and his driver. Obviously, neither of them heard us. Most importantly, they didn't expect us.

Where Is A Military Cop When You Need One?

"Now, stateside and at other locales, there would be military police (MPs) around directing traffic. We hadn't seen any MPs for months. We were probably about 75 yards back when they came out; we were big, heavy and moving quickly. Fortunately, we were slowing for the junction ahead.

"Our intercom was always locked open, and I heard Bratcher yell: 'You dumb ....! That **** doesn't even see us.'

"While we were in full tactical mode, as we were then, we were not buttoned up. The 90 (gun) was out front, rather than locked in the rear travel cradle. So, the fat muzzle brake was hanging right out there. I was hanging on to the .50 cal. gun, and we were all hanging out of our hatches.

"I told Ray to glide close, but be careful. We were right behind them with the gun muzzle looking like it was hanging right over their fancy spare tire. Suddenly, the jeep swerved a little back and forth, probably because the driver spotted us in his rearview mirror. He also must have made some noise, because his passenger, a major or a Lt. Colonel, looked back at us. From the look on his face, I swear to you he must have 'marked his laundry' right then.

"The driver jerked the wheel a little bit more, quickly pulled over, and stopped. The commander turned around the other way and looked up at me. I snapped a salute as we went by.

"Luckily, Ray was slowing down for him and the turn coming up. If he had not been, our 'rooster tail' of road dust would have given them and their jeep a new shade of tan.

Don't Practice Distracted Driving Habits

"A couple days later, word was passed down that all road convoy conventions should be strictly adhered to. Of course.

"We are in a hurry to get up there and definitely How Able [HA] to get out of there. So, as far as I'm concerned, there were probably three atomic cannons in Korea in 1953. There's most likely a buried set of fatigues marking the location."

And therein lies the whole crux of the story of "Annie" in

Korea. Lalley admitted he did not actually see the gun. Others said they had seen "something like it." In neither case did anyone actually see "Annie," and seeing something "like it" is not actually seeing it. That something like it might have been the 240mm cannon, which was a common error, as 2nd Lt. Bob Black explained.

Chapter 26
240? 280? What's 40mm Among Friends?

"The scientist is not responsible for the laws of nature. It is his job to find out how these laws operate. It is the scientist's job to find the ways in which these laws can serve the human will. However, it is not the scientist's job to determine whether a hydrogen bomb should be constructed, whether it should be used, or how it should be used. This responsibility rests with the American people and with their chosen representatives." Edward Teller

One of the problems with the cannons U.S. artillery units had in Korea was their penchant to blow up in the process of firing rounds. That was one of the worries the gun crew on "Annie" had in May 1953 during their first test. The threat was real, as U.S. Army Armament Officer Bob Black observed.

Black's responsibility was to find out why so many cannons were blowing themselves up in Korea during the war instead of blowing up enemy targets. The 280mm cannon was one of his particular concerns. Army brass simply did not want an atomic cannon to blow up in their faces, figuratively or literally. Neither did he.

Big Gun, Big Rumors

Black was familiar with both the 240mm and 280mm cannons. He opined that the troops in the field could not tell the difference, which accounted perhaps for the alleged sightings of "Annie" in Korea. That's not surprising, as William Russell alluded to.

He wrote that, "The atomic cannon…was basically a 'super' 240 mm gun that was the biggest in the arsenal and mounted on the German K5 railroad gun carriage." The troops' inability to distinguish between the 240 and 280 cannons was one of his main points in this extract of an article Black wrote for The Graybeards:

Why Are Our Guns Blowing Up?

By Bob Black

"I was a 2nd Lieutenant and Armament Officer in IX Corps, 17th D.S. Company, Chorwon/Kumwha Valley Area, in the Iron Triangle in 1952-53. I was with the 82nd Airborne at Ft. Bragg, NC when I received orders for Aberdeen Proving Ground, MD to attend an Armament Officer school. It seems that they had a problem with some of our guns in Korea blowing up and killing the gun crews.

"I grew up working in my father's garage, where I was introduced to being a machinist/mechanic. I had good knowledge of and could use calipers, micrometers, pull over gauges, etc. That came in handy, as all the men with whom I was assigned to the class at Aberdeen had similar machine/mechanic backgrounds.

"We studied all the guns, howitzers, mortars, and automatic crew-served armament weapons from A to Z there. The atomic cannon was one of the guns we came in contact with. It was 280mm bore, which was of a similar type to the German guns named "Leopold" or "Anzio Annie" 280mm Bruno class-railway guns or fixed position guns.

"All material we studied was classified. The huge gun was mounted on a cradle with two large tractors to propel it, one pulling in front and one pushing in the rear. It was big, awkward, extremely heavy, and difficult to move due to its weight and size. I was told that two guns were in Europe and none were in Korea. (Perhaps the people who saw it in Korea were looking at our largest gun there, a 240mm gun.)

"Anyway, I had a sixty-man armament platoon in that sector in Korea. We inspected any gun that needed it, and test fired any gun if there were questions about it. My platoon was scattered all over the firing lines. Our motto was 'Service to the line on the line.' I spent most of my time test firing guns that were hit in counter-battery fire or satisfying requests for service from the Field Artillery Battalions we were supporting.

"We had 300 guns and howitzers in our sector. The 105 was a good artillery piece. We had a ROK unit that tied the lanyard and was firing the gun automatically, believe it or not.

The guns were diverse. They included the 155mm howitzer, which put out a lot of iron on targets. The 8" howitzer ("Big 8") was the most accurate gun I had in the 424 Field Artillery Battalion. Then there were the 155 SP (Self Propelled) guns operated by the 937 Field Artillery Battalion.

"These guns, also known as 'The Arkansas Long Toms,' could really 'scoot and shoot.' They were loud and nasty, and covered a lot of ground. The 155 SPs comprised the bulk of the guns for which I was responsible.

"The 240mm guns were few and far between. A crane was required to assemble a gun from two carriers. One had the tube assembly; the other carried the trailer and cradle assemblies."

Dudley Middleton** explained that it was fairly easy to differentiate between the two cannons. "The atomic cannon was not a split tail, but was a gun inside a metal box similar to a railroad car. (The story from "Ski" Sherman presented in Chapter 1 has a very good description of the gun and its transporters.)

**Middleton was not in Korea during the war. He was a Korea War Defense Veteran, which was the term used for personnel who served in South Korea after 27 July 1953. Korean War veterans were those who had served during the actual fighting, technically 25 June 1950 to 31 January 1955.

During my 1958 tour of Korea, we fired close to 100 rounds of HE (High Explosive) from our battalion's six guns. This was a very accurate weapon, with a terrific muzzle velocity (MV). In fact, when we attempted to measure the MV using electronic rings, the first shot broke the rings. So, we continued shooting all of the guns with the allotted ammunition. I later did a 'fall of shot' calibration for the battalion."

Middleton's observation about recognizing the difference between the cannons sounds simple. The problem was that the 240mm and 280mm cannons did not operate in the same areas in Korea—again assuming that "Annie" was ever there. As William Russell observed, "Only those soldiers who were crew members of the gun(s) in Korea know for certain."

One question arises from Black's story: how could non-artillery troops distinguish a 105 Howitzer from a 155 Howitzer from a 240mm? Many of them knew that there were self-propelled 155s, and that the 240 mms were also self-propelled. So, when news of the 280mm cannon arrived in Korea it was only natural that troops jumped to the conclusion that the gun had arrived along with it. (News was hardly delivered in real time back then.) Military and political leaders hoped that those rumors would reach enemy commanders as well. But, news and reality were two different things.

The Rumors Were Good For Troop Morale

Andrew Antippas corroborated Black's comments about people mistaking the 240mm guns then in Korea for the 280mm atomic cannons. "I clearly recollect that there was talk in the spring of 1953 in my battalion about atomic cannon being in country following the tests in Nevada in May," he said.

"The scuttlebutt was that President Eisenhower was fed up with delays in signing the Armistice and was letting it be known that he was prepared to use atomic weapons if the communists didn't stop their delaying tactics. As one who would have been intimately involved in any resumption of attacks to the north, I was interested.

"I had some knowledge of the atomic cannon, since I had been on a tour of the Watertown Arsenal in 1951 as a member of the American Ordnance Association, and was shown the completed barrel of the atomic cannon. I still recall the enormous size of the square breechblock, the size of a small refrigerator, on the 280mm barrel," he concluded.

No doubt the troops found the rumored presence of "Annie" a great topic of conversation. By the spring of 1953 they were bogged down in a mountain-to-mountain type of warfare that was tedious and costly to both sides in terms of injuries and deaths. They would have welcomed the presence of a cannon that could fire nuclear weapons—or any other nuclear weapons, for that matter. But rumors were all they had to discuss. Like everybody else, they had to wait until 1958 for nuclear weapons to arrive in Korea, long after all of them had gone home.

They may have gone home, but the hostilities continued. The threat of nuclear weapons stayed on the negotiating table for another five years, until the UN Command finally acted on it. After all the discussions of nuclear weapon use during the Korean War, none of them actually arrived in the country until 4-1/2 years after the 1953 cease fire—and then not without a lot of controversy.

What Good Is "Annie" Without "Honest John?"

Evidence suggests that in actuality the first 280mm cannon did not arrive in Korea until January 2, 1958. The U.S. Army sent along with it some Honest John nuclear-capable missiles which, like "Annie," could carry conventional or nuclear warheads. According to the Air Force Space and Missile Museum, "Because of its size and simplicity, Honest John had considerably more battlefield mobility than conventional artillery. Just one nuclear warhead carried aboard an Honest John could deliver the destructive power of hundreds of artillery shells." Thus, "Annie" became a part of a formidable team in Korea, albeit it a few years later than its rumored arrival.

Even though "Annie"—the original "Annie" that had fired the nuclear round at Frenchman Flat—and two siblings reached Korea, they never fired anything but conventional shells, and not many of those, according to a crew member.

Dudley Middleton was stationed in Camp Barbara, Korea, several miles north of Camp Casey, from January 1958 to February 1959. He revealed that he directed the fire of our two 280mm guns. "We only fired HE, and only into the Imjin River firing range. We never fired at any enemy," he said.

He noted that one of the guns in "A" Battery, which was stationed in Munsan-ni, had been the gun which was tested in the U.S. desert with an atomic shell just to be sure it worked. "That gun was called 'Atomic Annie,'" Middleton observed. "The last I knew it was sitting on the ground at the Fort Sill [OK] Museum."

"C" Battery had two guns," he continued. "They were stationed at Uijonbu. In all, our battalion had six guns. I understand there was a battalion of 280s stationed in Germany at about the same time.

"Our guns had been stationed in Okinawa prior to my assignment to "B" Battery. I am sure they were used as a deterrent to North Korean military ambitions. However, sometime after I left Korea, both parties signed an agreement to remove all atomic weapons from Korea. That was when 'Atomic Annie' went to the museum," he concluded.

One Army veteran recalls seeing a 280mm cannon in Japan. W.R. Ames said, "In 1956 I was with the 1st Cavalry at Camp Drake, outside Tokyo. I was in 15th QM (Quartermaster), which included the division transportation. Our motor pool was just outside Camp Drake. There was a 280 atomic cannon in a secured motor pool next to us. It was there for only a month or so.

"The cannon was heavily guarded 24 hours a day by 2 American guards walking inside and 2 Japanese guards walking outside."

Ole, The Matadors Arrive

Not to be outdone, the U.S. Air Force sent its own nuclear weapons in Korea a year later. It announced on December 16, 1958 that it had stationed a squadron of nuclear-tipped Matador cruise missiles in Korea—nine years after they were first tested.

The Air Force had an identity crisis with the Matador, the first operational surface-to-surface cruise missile the U.S. built. The nuclear-warhead equipped Matador featured a radio link that allowed in-flight course corrections, which enhanced the missile's accuracy over ranges of about 600 miles. Matadors were self-propelled by a small turbojet engine. So, the Air Force labeled them bombers.

Later, the Air Force changed the Matador's designation to tactical missiles. Whatever they were called, they didn't arrive in Korea until late 1958, where they remained until 1962, when they were taken out of service.

Hey, They Violated The Truce First

There was some question regarding the legitimacy of the installation of nuclear weapons in Korea. UN officials brushed them aside. They justified the Matadors' presence in a simple statement in a December 1958 press release, saying succinctly that their introduction "is in accordance with the spirit and intent of the UN position stated at the 75th MAC (Military Armistice Commission) meeting on June 21, 1957."

The UN justified its introduction of Matadors by claiming that "the communists, had built up, in violation of the armistice agreement, a vast military capability which upset the military balance the armistice was intended to preserve." In light of that development, the UN stated, and to rectify the imbalance, "the sub-paragraph of the armistice agreement which prohibited an arms build-up was being waived to permit restoration of a relative military balance."

The communists were not pleased with the UN's position. They protested. The UN waved off the protests as easily as they had waived the "sub-paragraph" mentioned above. UN representatives told the communists bluntly that the Matadors would stay in Korea.

U.S. Navy Rear Admiral Ira H. Nunn informed the communists' representatives at the Panmunjom peace talks that "the present situation was the result of their own introduction of modern weapons in violation of the armistice agreement." Just to appease them, he added that the new weapons were being introduced for defensive purposes only. The sparring went on—and the Matadors stayed until 1962, when they were retired from active duty.

A year later, "Annie" and all its siblings were taken out of service. "Annie" had done its job and, like all advanced weapons, gave way to more state-of-the-art nuclear weapons. That ended the career of one of the most influential pieces of artillery in Korea, which accomplished its purpose without ever firing a shot in anger.

Chapter 27
Not Much On Which To Build A Reputation

"In no other type of warfare does the advantage lie so heavily with the aggressor." James Franck

The fighting portion of the Korean War ceased at 10 p.m. on 27 July 1953 when the two sides signed a truce. Despite the fact that there was no proclaimed "official" victor in the war, there was one clear-cut winner: South Korea, which has grown from a poverty-stricken half of a divided nation to one of the top economies in the world. Conversely, its neighbor North Korea is one of the poorest. But, could the UN have earned a total victory if the U.S. had used nuclear weapons?

To guess would be useless speculation. The bottom line is that in the long run nobody in a leadership position was willing to pull the trigger on a nuclear weapon. The two sides reached an accord to end the fighting before anyone could—or would.

William Russell sums up the arguments about "Annie," nuclear weapons, and threats:

"Much has been written about the administration's debate in the winter and spring of 1953 about whether to expand (or end) the war with nuclear weapons and compel the Communists to accept an armistice. By May of 1953 it appears that the decision to use nuclear weapons was on the planning table.

"Based on this, it would seem improbable that the atomic cannon was deployed before the armistice. In fact, according to reports, the atomic cannon and the Honest John nuclear-capable missiles were introduced into Korea in January 1958, when the

Reds had begun their massive military build-up in violation of the ceasefire terms.

"The Communists came to the conference table in June 1951, but it was not likely under a nuclear threat as has been suggested. The atomic gun was still in development at that time. And, as everyone knows, negotiations went on for the next two years. In December 1952, when I was serving in Korea, Eisenhower visited the war zone and promised to end the war. How? Did he have the atomic cannon in mind?

"As has been reported, some have suggested he did, but the atomic cannon was not emplaced in Korea at that time. It was not deployed until January of 1958, according to information on the internet. However, only those soldiers who were crew members of the gun(s) in Korea know for certain.

"So far as ending the war, it was seven months after Eisenhower became president and eight months after he came to Korea with the promise to end the war that a ceasefire was signed by both parties. In Eisenhower's defense, however, he didn't say when. Whether or not the Communists were influenced by the threat of nuclear action is debatable." Russell was not alone in that belief.

Historian David Holloway, who compiled a history of the early years of the Soviet nuclear weapons program, said there is little evidence to suggest that the threat of the U.S.'s nuclear weapons compelled Russia to deviate from its normal policies regarding foreign relations. That was the case throughout the Korean War and beyond, right to the present day.

Today, the two Korean War adversaries remain technically in a state of war. Many of the principal countries engaged in the war have nuclear weapons at their disposal. The world has witnessed fits and starts regarding the use of nuclear weapons since 1953, but still no one has employed them to date—and there is still no central control over their use.

Some of the weapons have come and gone as technology renders them outdated. "Atomic Annie" is a prime example of how quickly technology changes. By 1963 the once revolutionary nuclear cannon was retired and relegated to being a museum piece.

Korean War veteran Richard Santora recalls seeing a version of "Annie" years after the war, and recalls that it was fired for visitors: "In the past years, around May, Aberdeen Proving Grounds opened many of its facilities to the public. One of the most popular was the fire demonstration at the 'range front.'"

"Weapons systems, tanks, antitank vehicles, small arms, troop maneuvers, etc. were demonstrated, using blank and/or live ammo. One of the big features of the show was setting up and firing the "atomic cannon," using a conventional projectile, which impacted several thousand yards down range. It was billed as an atomic cannon, and may be similar to the one claimed to have been sighted in Korea."

Santora's use of the word "claimed" sums up the arguments about its use that continue today, 60+ years, after "Annie" became a weapon in Korea. Whether the claims of its use there are true or not, "Annie" still has the honor of being the only artillery piece that ever actually fired a nuclear shell. That may not seem like much on which to base a reputation, but "Annie's" mere existence may have been enough to bring the Communists to the peace talks table.

May have? Again, gauging the influence of a weapon like "Annie" on the Communists' ultimate decision to seek a cease fire is speculation. The troops—the men and women who benefited most from that decision—like to think it helped. That's the biggest benefit a weapon like "Annie" can provide in times of war, even if it never did fire a shot in anger. The same holds true for the other nuclear weapons never used in Korea.

In the final analysis, it is difficult to determine just how influential the threats of nuclear weapons use were in the outcome of the Korean War. Truman and Eisenhower waxed hot and cold on their use, and their military leaders did the same. No one was sure if their use would have any beneficial result on the conduct of the war. Consequently, generals, presidents, secretaries of state, war, and other departments restricted the topic of nuclear weapons to discussions. Historians have reacted accordingly.

Historians argue among themselves as to whether Truman and Eisenhower actually threatened to employ nuclear weapons in Korea or simply implied that they would. The argument boils down to semantics. The fact is that actual threats or implied threats aside, no one authorized the use of nuclear weapons. The Communists finally came to the peace talks table regardless, and the two sides worked out a cessation of hostilities. Would they have had that opportunity if the UN had employed nuclear weapons in Korea? That is a question with no answer.

In retrospect, the Korean War can be considered the second nuclear war even though nuclear weapons were not used. The specter of their use hung over the combatants almost from the beginning of the war. "Atomic Annie," "Honest John," "Matadors" and other nuclear weapons spent their hour upon the stage and then were heard no more. The world was saved from a second use of nuclear weapons simply because no political or military leader was willing to make the decision to employ them--or has been since.

Perhaps that is the most significant outcome of the Korean War. U.S. and UN political and military leaders proved that it does not require the use of nuclear weapons to resolve differences among nations. Their only value may be psychological. Hopefully, that message can be applied to war in general for the salvation of future generations.

APPENDIX A
UN Resolution 82

The Security Council,

Recalling the finding of the General Assembly in its resolution 293 (IV) of 21 October 1949 that the Government of the Republic of Korea is a lawfully established government having effective control and jurisdiction over that Part of Korea where the United Nations Temporary Commission on Korea was able to observe and consult and in which the great majority of the people of Korea reside; that this Government is based on elections which were a valid expression of the free will of the electorate of that part of Korea and which were observed by the Temporary Commission, and that this is the only such Government in Korea,

Mindful of the concern expressed by the General Assembly in its resolutions 195 (III) of 12 December 1948 and 293 (IV) of 21 October 1949 about the consequences which might follow unless Member States refrained from acts derogatory to the results sought to be achieved by the United Nations in bringing about the complete independence and unity of Korea; and the concern expressed that the situation described by the United Nations Commission on Korea in its report menaces the safety and well-being of the Republic of Korea and of the people of Korea and might lead to open military conflict there,

Noting with grave concern the armed attack on the Republic of Korea by forces from North Korea,

Determines that this action constitutes a breach of the peace; and

I

    Calls for the immediate cessation of hostilities;

    Calls upon the authorities in North Korea to withdraw forthwith their armed forces to the 38th parallel;

II

    Requests the United Nations Commission on Korea:
- a. To communicate its fully considered recommendations on the situation with the least possible delay;
- b. To observe the withdrawal of North Korean forces to the 38th parallel;
- c. To keep the Security Council informed on the execution of this resolution:

III

    Calls upon all Member States to render every assistance to the United Nations in the execution of this resolution and. to refrain from giving assistance to the North Korean authorities.

    −text of UN Security Council Resolution 82[16]

APPENDIX B
UN Resolution 83

The Security Council,

Having determined that the armed attack upon the Republic of Korea by forces from North Korea constitutes a breach of the peace,

Having called for an immediate cessation of hostilities,

Having called upon the authorities in North Korea to withdraw forthwith their armed forces to the 38th parallel,

Having noted from the report of the United Nations Commission on Korea1 that the authorities in North Korea have neither ceased hostilities nor withdrawn their armed forces to the 38th parallel, and that urgent military measures are required to restore international peace and security,

Having noted the appeal from the Republic of Korea to the United Nations for immediate and effective steps to secure peace and security,

Recommends that the Members of the United Nations furnish such assistance to the Republic of Korea as may be necessary to repel the armed attack and to restore international peace and security in the area.

Adopted at the 474th meeting
by 7 votes to 1 (Yugoslavia).2

APPENDIX C

Remembering general's crash
by Mark Wilderman
60th Air Mobility Wing History Office
8/10/2011 - TRAVIS AIR FORCE BASE, Calif. -- On August 5, 1950, the worst disaster in the history of Travis occurred. Brig. Gen. Robert Travis, commander of the 5th Strategic Reconnaissance Wing and the 9th Bombardment Wing at Fairfield-Suisun Air Force Base died at age 46 in the crash of the Boeing B-29 "Superfortress" in which he was traveling. The aircraft crashed five minutes after a nighttime takeoff at Fairfield-Suisun AFB as part of a 15-ship deployment to the Pacific just after the beginning of the Korean Conflict.

The crash killed 11 other crewmembers and passengers aboard the B-29 and seven people on the ground, including base firefighters and volunteers attempting to rescue the crew. In addition, 49 injured people were admitted to the hospital and 124 others received minor injuries. Eight of the B-29's crew and passengers (including both pilots) survived the crash.

According to the Air Force accident report, the cause of the crash was a number two engine propeller malfunction at liftoff, combined with the failure of the landing gear to retract, causing the aircraft to be unable to climb from an altitude of 200 feet. The aircraft's left wing struck the ground at 120 mph as the pilots attempted to make a 180-degree turn to the right back towards the base for a landing. Approximately 20 minutes after the crash and fuel fire, the highly-explosive filler in the aircraft's bomb casing ignited, resulting in a blast that severely damaged a base trailer park near the main gate and was clearly heard 30 miles away in

Vallejo.

The bomb explosion killed five base firefighters, Pvt. Emile Bender, Jr., Pvt. Edward Goins, Cpl. Doyle Hanstead, Staff Sgt. John McCollum, and Pfc. William Vetter. Two volunteers also lost their lives, Sergeant Paul Ramoneda of food services, who died heroically while attempting to rescue twelve passengers and crew trapped aboard the burning B-29, and Private John Boyles. Sergeant Ramoneda, a decorated World War II Marine Corps veteran, was posthumously awarded the Soldier's Medal, the Purple Heart and the Cheney Medal, (awarded annually since 1927 for an act of valor, extreme fortitude or self-sacrifice in a humanitarian interest performed in connection with aircraft. The Travis Airman Leadership School was later named in his honor.

General Travis saw combat action in World War II as commander of the 41st Combat Wing in England, personally leading 35 combat missions over Nazi-occupied Europe.

On Oct. 2, 1950, an Air Force special order officially renamed Fairfield-Suisun AFB in honor of the fallen commander. The formal Travis AFB dedication ceremony was held on April 20, 1951, presided over by the governor of California, Earl G. Warren, many prominent dignitaries and the Travis family. The dedication ceremony included a parade and a flyover of the massive Convair B-36 "Peacemaker" heavy bomber.

Comments

3/16/2014 12:19:33 PM ET

According to Eric Schlosser in Command and Control on pages 168 and 169, the B-29 was headed to Guam with a Mark 4 atomic bomb under urgent secret orders from President Truman for possible use against North Korea. The nuclear core was being flown in a separate plane so was not involved in the crash.

David C Hall, Lopez Island WA

APPENDIX D

SCB-27 modernization of Essex/Ticonderoga class aircraft carriers

Between 1947 and 1955, fifteen Essex and Ticonderoga class aircraft carriers were thoroughly modernized. The impending arrival of high-performance jet aircraft and nuclear-armed heavy attack bombers had rendered these still rather new ships almost incapable of executing their most vital missions, while the post-World War II financial climate precluded building replacements. Accordingly, a reconstruction program began in Fiscal Year 1948, with the incomplete Oriskany as the prototype. Two more ships were converted the next year, three in FY 1950 and then, with the Cold War in full bloom, nine more Fiscal Years 1951 to 1953.

Designated SCB-27, the modernization was very extensive, requiring some two years for each carrier. To handle much heavier, faster aircraft, flight deck structure was massively reinforced. Stronger elevators, much more powerful catapults, and new arresting gear was installed. The original four twin 5"/38 gun mounts were removed. The new five-inch gun battery consisted of eight weapons, two on each quarter beside the flight deck. Twin 3"/50 gun mounts replaced the 40mm guns, offering much greater effectiveness through the use of proximity-fuzed ammunition.

A distinctive new feature was a taller, shorter island. To better protect aircrews, ready rooms were moved to below the armored hangar deck, with a large escalator on the starboard side amidships to move airmen up to the flight deck. Internally, aviation gasoline storage was increased by nearly half and its pumping capacity enhanced. Also improved were electrical generating power, fire protection, and weapons stowage and handling facilities. All this added considerable weight: displacement increased by some twenty percent. Blisters were fitted to the hull sides to compensate, widening waterline beam by eight to ten feet. The ships also sat lower in the water, and maximum speed was slightly diminished.

The modernized ships came in two flavors, the first nine (SCB-27A) having a pair of H 8 hydraulic catapults, the most powerful available in the late '40s. The final six received the SCB-27C update, with much more potent steam catapults, one of two early 1950s British developments that greatly improved aircraft carrier potential. These six were somewhat heavier, and wider, than their sisters. While still in the shipyards, three of the SCB-27Cs were further modified under the SCB-125 project, receiving the second British concept, the angled flight deck, plus an enclosed "hurricane bow" and other improvements. These features were so valuable that they were soon back-fitted to all but one (Lake Champlain) of the other SCB-27 ships. The fourteen fully modernized units were the "journeymen" aviation ships of the late 1950s and 1960s, providing the Navy with much of its attack aircraft carrier (CVA) force and, ultimately, all its anti-submarine warfare support aircraft carriers (CVS).

The SCB-27 program involved rebuilding fifteen ships, three of which were given a combined SCB-27 and SCB-125 modernization. (Lake Champlain was reconstructed to SCB-27A design by the Norfolk Naval Shipyard. Work began in August 1950; the ship was recommissioned in September 1952.)

Source:

http://www.patriotfiles.com/index.php?name=Sections&req=viewarticle&artid=3601&allpages=1&theme=Printer

APPENDIX E

AJ Savage Bomber

The AJ Savage was the first U.S. bomber designed especially to carry the atomic bomb. It was North American's first attack bomber for the U.S. Navy and was designed shortly after the end of World War II. It was a large twin-engine Heavy Attack aircraft for the Navy, as big as the Air Force medium bombers of the time, such as the B-45 Tornado.

In those early years of jet aircraft development, manufacturers were exploring ways to provide power using piston engines and a jet engine on the same airframe. The AJ-1 attack bomber used two 2,400 horsepower piston engines to power four-bladed propellers for long-range cruise. It then fired a 4,600-pound-thrust turbojet engine for extra speed over the target.

The AJ Savage had a crew of three and a single tail unit. Its folding wings allowed it to be stored on an aircraft carrier. After building three XAJ-1 prototypes and a static test model, North American began delivering the AJ-1.

The Savage entered service in September 1949 and carrier operations began in April 1950 on the USS Coral Sea. North American built more than 140 in the series. Later, some AJ models were converted into aerial tankers. Others, the AJ-2Ps, with a modified radome (a contraction of radar and dome: a structural, weatherproof enclosure that protects a microwave or radar antenna, and through which radio waves can pass) carried 18 cameras. Their night shots were illuminated by a photo-flash unit in the fuselage. These models were standard equipment for the Navy heavy photographic squadrons until the early 1960s.

Specifications:

First flight: July 3, 1948
Span: 75 feet 2 inches
Length: 63 feet 1 inches
Gross weight: 52,862 pounds
Power plant: Two 2,400-horsepower Pratt & Whitney R-2800-44W piston engines, 4,600-pound thrust Allison J33-A-10 turbojet engine
Max speed: 471 mph (all engines)
Crew: Three
Range: 1,630 miles
Source: Boeing History, http://www.boeing.com/history/bna/ajsavage.htm

APPENDIX F

Atomic Annie on the move
  September 16, 2010
  By Mr. Jeff Crawley (IMCOM)
  Directorate of Logistics employees prepare to move the M65 Atomic Cannon, "Atomic Annie" Sept. 14 from the Fort Sill Cannon walk. The atomic cannon is just one of the field artillery pieces that will be moved to the new Artillery Park next to the Fort Sill Museum.
   FORT SILL, Okla.--The M65 Atomic Cannon "Atomic Annie" at Fort Sill since 1964 is moving. "Atomic Annie" moved Sept. 14 from the Cannon Walk at Geronimo and Randolph roads to a staging area here where it will get a new coat of paint. After refurbishing, its home will be at the new Artillery Park next to the Fort Sill Field Artillery Museum. The towed artillery piece, which fired a 280mm nuclear shell up to 20 miles, was a deterrent during the Cold War, but its technology was quickly overtaken by missiles and rockets, said Towana Spivey, Fort Sill Museum director and curator. "Atomic Annie," originally nicknamed "Able Annie," was one of 20 guns produced in the early 1950s. In May 1953, two of the guns from Fort Sill were taken to Frenchman Flat test site in Nevada, Spivey said. Only one of the guns, "Able Annie" was fired in the nuclear tests called Knothole, while "Sad Sack" served as a backup. "The gun fired a 15-kiloton projectile equal to the bomb that was dropped on Hiroshima," Spivey said. Immediately after the firing "Able Annie" was rechristened "Atomic Annie." During the transport of the cannons back to Oklahoma, the identity of them somehow got switched. "Sad Sack," thought to be "Atomic Annie," made it to Fort Sill. The true

"Atomic Annie" was sent to an operational unit. During the 10th anniversary commemorating the firing of "Atomic Annie," it was noticed through serial number identification that the gun was really "Sad Sack," Spivey said. So what happened to "Atomic Annie'" The exact locations of atomic guns were classified and they were scattered throughout Europe and Southeast Asia, he said. A worldwide search began for "Atomic Annie," who was now being referred to as "AWOL Annie," Spivey said. "Only a very few people knew where it was." The gun was tracked to Houston, New York City and eventually found on duty at a remote site in Germany.

In 1964, while the gun was being retrieved in Germany, "Atomic Annie" and its tractor slid off a mountain road killing two Soldiers and wrecking the prime movers, Spivey said. That same year, "Atomic Annie" no longer AWOL replaced the imposter at Fort Sill, he said. "Sad Sack" was given to the Smithsonian museums and is part of its collection.

SOURCES

PRIMARY SOURCES

W. R. Ames
Andrew Antippas
Richard E. Bailey
Frank Barron
Raymond Bistline
Charles W. Boyce
Stan Britton, Sr.
Lindy Constantino
Peter W. Cuthbert
Thaddeus Czarnowski
Edgar E. Danley
Ralph Delaney, Jr.
Bob Hall, Bellingham
William H. ("Bill") Harris
Jack L. Hatchitt
James C. Henderson
Rolland K. Hindsley
Frank A. Imparato
Richard K. Jenkins
Roland Jennings
Albert A. Kamishlian
Dick Larrowe

Jim Low
George H. McKenzie, Jr.
Dudley Middleton
Dennis Mueller
James L. Murphy
Robert A. Palmrose
George Parks
Donald L. Parrott
Dick Payne
Arles W. Pease
Marvin Reed
Robert W. Robinson
William Russell
William H. Sanford
Richard Santora
Wayne A. Schild
Marvin H. Schafer
Melvin Schriefer
George Sherman
John W. Sonley
Paul Theiring
Leon S. Wozniak

SECONDARY SOURCES

BOOKS

Ambrose, Stephen E. Eisenhower: Soldier And President. New York: Touchstone. 1990.
Blair, Clay. The Forgotten War. New York: Times Books. 1987.
Campbell, John Martin. Slinging the Bull in Korea. Albuquerque, NM, University of New Mexico Press. 2010.
Catchpole, Brian. The Korean War. New York: Carroll & Graf Publishers, Inc. 2000.
Channon, Robert I. Airborne Quarterly, Winter 2014.
Cumings, Bruce. The Korean War. New York. The Modern Library. 2010.
Fehrenbach, T.R. This Kind Of War: The Classic Korean War History. Dulles, VA: Potomac Books, 1963.
Halberstam, David. The Coldest Winter. New York, Hyperion. 2007.
Haruki, Wada. The Korean War. Lanham, MD, Rowman & Littlefield. 2014.
Korda, Michael. Ike: An American Hero. New York: Harper Collins. 2007
Leckie, Robert. The War In Korea 1950-1953. New York, Random House. 1963.
Lee, Hubert Hojae. My Journey To America. Washingtonville, NY. The Spear Printing Co. 2010.
McCullough, David. Truman. Touchstone. New York, 1992.
Miller, John Jr. et al, Korea 1951-53. Center of Military History: Department of the Army. 1989.
Sharp, Arthur G. The Everything Theodore Roosevelt Book. Avon, MA, Adams Media. 2011.
Smith, Jean Edward. Eisenhower In War And Peace. New York. Random House. 2012.
Truman, Harry S. Memoirs, Vol. II. Years of Trial and Hope. Doubleday Co. Garden City, NY, 1956.
Truman, Margaret. Harry S. Truman. William Morrow & Co., Inc. New York, 1973.
Wicker, Tom. Dwight D. Eisenhower. Times Books: New

York. 2002.

MISCELLANEOUS

1946 "Crossroads" Nuclear Test Report 25th Division, U.S. Army. http://www.atomicvetkin.com/crossroads.html
    280mm Atomic Annie Artillery. http://olive-drab.com/idphoto/id_photos_atomic.php
    http://www.history.army.mil/documents/Korea/25id-KW-IP.htm
    Arlington National Cemetery.
    http://www.arlingtoncemetery.net/ridgway.htm
    Atomic Annie On The Move.
    http://www.army.mil/article/45311/atomic-annie-on-the-move/
    "Dien Bien Phu: Did the US offer France an A-bomb?" BBC News Magazine, 5 May 2014.
    Buck, Alice. The Atomic Energy Commission. U.S. Department of Energy.
    http://energy.gov/sites/prod/files/AEC%20History.pdf
    _____ "Why Did Truman Really Fire MacArthur? ... The Obscure History of Nuclear Weapons and the Korean War Provides the Answer."
    http://www.historynewsnetwork.org/article/9245#sthash.0BtE38GT.dpuf
    Defense Threat Reduction Agency (VA) http://www.dtra.mil/documents/ntpr/factsheets/Hiroshima_and_Nagasaki_Occupation_Forces.pdf
    Eger, Chris. The US Army's Nuclear Cannon: Atomic Annie (VIDEO). http://www.guns.com/2013/08/15/atomic-annie-the-us-armys-nuclear-cannon/ , 8/15/13.
    Gwertzman, Bernard. "U.S. Papers Tell Of '53 Policy To Use A-Bomb In Korea," June 8, 1984, New York Times. - http://www.nytimes.com/1984/06/08/world/us-papers-tell-of-53-policy-to-use-a-bomb-in-korea.html
    Korea Reborn: A Grateful Nation. RememberMyService.com
    http://www.koreanwar60.com/biographies-hoyt-s-

vandenberg

International Control of Atomic Energy, 1946: http://www.foia.cia.gov/sites/default/files/document_conversions/50/Report_on_the_International_Control_of_Atomic_Energy_16_Mar_1946.PDF

Lamont, James. The Atomic Cannon: It was fired only once, but it helped end a war http://www.chymist.com/The%20Atomic%20Cannon.pdf

MacArthur to Martin. http://ww2db.com/doc.php?q=408

Mortality of Veteran Participants in the CROSSROADS Nuclear Test. Institute of Medicine. http://www.iom.edu/Reports/1996/Mortality-of-Veteran-Participants-in-the-CROSSROADS-Nuclear-Test.aspx.

New York Times, "Text of Accounts by Lucas and Considine on Interviews With MacArthur in 1954," The New York Times, April 9, 1964, pg. 16.

"Nukes in the Taiwan Crisis." http://fas.org/blogs/security/2008/05/nukes-in-the-taiwan-crisis/

"Operation Vulture," http://en.wikipedia.org/wiki/Operation_Vulture

Oppenheimer, J. Robert. "Atomic Weapons and American Policy." Foreign Affairs, July 1953. Vol. 31, #4.

http://www.thenakedscientists.com/forum/index.php?topic=7801.0

SBC-27 program. http://www.patriotfiles.com/index.php?name=Sections&req=viewarticle&artid=3601&allpages=1&theme=Printer

"Remembering General's Crash," http://www.travis.af.mil/news/story.asp?id=123267480U.S.

Department of State: Milestones: 1945-1952, Korean War 1950-1953.

http://www.strategic-air-command.com/weapons/nuclear_bomb_chart.htm

The Crash of the B-29 on Travis AFB, CA http://www.check-six.com/Crash_Sites/Travis_B-29_crash_site.htm

"UNC In Korea Gets Matador Missiles," Pacific Stars And Stripes, Thursday, December 18, 1958, Vol. 14, No. 351.

http://www.tacmissileers.org/korea-gets-matador-missiles/
http://www.afspacemuseum.org/displays/HonestJohn/
http://www.veteransforpeace.org/files/2113/3348/4164/korea_timeline.pdf
http://history.state.gov/milestones/1945-1952/korean-war-2
http://eisenhower.archives.gov/research/online_documents/korean_war/I_Shall_Go_To_Korea_1952_10_24.pdf (Eisenhower's Detroit speech)
http://academic.mu.edu/meissnerd/eisenhower.html
The Bikini Atoll Survey "Operation Crossroads," 1946-47.
http://www.mnh.si.edu/onehundredyears/expeditions/bikini.html
The US Army's Nuclear Cannon: Atomic Annie (Before Its News)
http://beforeitsnews.com/alternative/2013/08/the-us-armys-nuclear-cannon-atomic-annie-video-2737870.html
http://en.wikipedia.org/wiki/MGM-1_Matador
NOTE: For an extensive list of Korean War histories and publications, see http://www.koreanwar60.com/publications

AUTHOR BIO

Arthur G. Sharp, who served four years with the U.S. Marine Corps, is the author of 2,500 articles and 14 books on various topics. His most recent book, The Siege of LZ Kate, relates the heroic efforts of a group of U.S. Army artillerymen to escape a siege enforced by 4,500 North Vietnamese soldiers and extols the extraordinary leadership efforts of two Green Berets in the operation.

Sharp is the editor of The Graybeards, the Korean War Veterans Association's 80-page bimonthly magazine, and The Old Breed News, the First Marine Division Association's quarterly 32-page magazine. He holds a B.A. and an M.A. in American history.

He holds B.A. and M.A. degrees in history from the University of Hartford and Trinity College respectively. Sharp, a native of Connecticut, currently resides in Sun City Center, Florida.